REGIONAL TECTONICS, NEW ENGLAND AND QUEBEC

The Geological Society of America, Inc.
Memoir 120

Regional Tectonic Synthesis of Northwestern New England and Adjacent Quebec

Wallace M. Cady
U.S. Geological Survey
Denver, Colorado

1969

Publication Authorized by the Director, U.S. Geological Survey
Library of Congress Catalog Card Number 77–98020
S.B.N. 8137–1120–7

Published by
THE GEOLOGICAL SOCIETY OF AMERICA, INC.
Colorado Building, P.O. Box 1719
Boulder, Colorado 80302

Printed in The United States of America

The printing of this volume
has been made possible through the bequest of
Richard Alexander Fullerton Penrose, Jr.,
and is partially supported by a grant from
The National Science Foundation.

Preface

This study is an outgrowth of regional geologic mapping and investigations of talc and asbestos deposits conducted in Vermont by the U.S. Geological Survey. The author is especially indebted to Arden L. Albee and Alfred H. Chidester, his associates in this project, whose criticism of the manuscript of this paper and of related official reports, has helped guide the orientation of the study. Drafts of the manuscript have also been reviewed by Professors Brewster Baldwin and Peter Coney of Middlebury College, Professors R. Y. Lamarche of the University of Sherbrooke, F. F. Osborne of Laval University, P. H. Osberg of the University of Maine, and John L. Rosenfeld of the University of California at Los Angeles; by Dr. Pierre St. Julien of the Quebec Department of Natural Resources, Dr. M. J. Rickard of Australian National University, and by James Gilluly, R. H. Moench, L. R. Page and R. E. Zartman of the U.S. Geological Survey. The author is indebted to the foregoing, and to R. H. Jahns of Stanford University and W. S. White of the U.S. Geological Survey for counseling in the field, for contributions of unpublished information, and for many other suggestions in times past.

Dr. J. R. Béland and Dr. F. F. Osborne and their colleagues of the Quebec Department of Natural Resources have kindly shown the author over much of the ground between the Vermont-Quebec boundary and the Gaspé Peninsula. Dr. Charles G. Doll, Vermont State Geologist, and his associates have provided abundant information well in advance of publication, which is now about complete. Professors M. P. Billings and J. B. Thompson, Jr., of Harvard University, have maintained a continued interest in the U.S. Geological Survey's work in Vermont, in support both of the regular investigations of the project and of such by-products as the present study. The author is much indebted to E-an Zen of the U.S. Geological Survey for a continuing correspondence concerning the relationships of the rocks of the Taconic Range.

Others of the author's colleagues who have contributed information, suggestions, and criticism as well as their interest and hospitality, and some of whom are members of the organizations above mentioned, include C. A. Anderson, Alex Baer, G. W. Bain, Robert Balk, H. M. Bannerman, E. S. Barghoorn, F. W. Benoit, M. B′rubé, V. H. Booth, A. J. Boucot, W. F. Brace, R. L. Brown, Jr., Josiah Bridge, W. H. Bucher, W. S. Burbank, E. N. Cameron, F. C. Canney, P. H. Chang, R. A. Christman, T. H. Clark, H. C. Cooke, G. W. Crosby,

R. V. Cushman, J. G. Dennis, H. S. de Römer, W. H. Diment, H. R. Dixon, F. A. Donath, Gilles Duquette, P. R. Eakins, G. P. Eaton, E. H. Ern, Jr., Henry Faul, M. M. Fitzpatrick, Phillip Fowler, B. K. Goodwin, W. A. Gorman, J. C. Green, P. E. Grenier, J. T. Hack, J. B. Hadley, L. M. Hall, Warren Hamilton, B. B. Hanshaw, N. L. Hatch, Jr., H. E. Hawkes, Jr., Norman Herz, B. F. Howell, A. L. Howland, Claude Hubert, W. P. Irwin, E. C. Jacobs, E. P. Kaiser, Marshall Kay, Arthur Keith, P. B. King, E. B. Knopf, R. H. Konig, P. J. Lespérance, J. B. Lyons, J. A. MacFadyen, Jr., R. A. Marleau, J. F. Murphy, V. R. Murthy, P. B. Myers, Jr., R. S. Naylor, J. D. Obradovich, K. L. Pierce, L. B. Platt, D. W. Rankin, R. B. Raup, P. E. Raymond, P. H. Riordon, John Rodgers, R. J. Ross, Jr., B. M. Schmidt, D. T. Secor, Jr., A. B. Shaw, Q. D. Singewald, J. W. Skehan, G. L. Snyder, C. W. Stearn, S. W. Stone, T. P. Thayer, George Theokritoff, Barry Voight, M. S. Walton, C. W. Welby, R. R. Wheeler, D. E. White, H. B. Whittington, D. R. Wones, and B. G. Woodland.

A. C. Christiansen, Marjorie MacLachlan, and E. K. Stoll of the Technical Reports Unit of the Geological Survey edited the illustrations, stratigraphic nomenclature, and text.

Contents

FIGURES

PLATES

TABLES

Abstract

The northwestern New England and adjacent Quebec region is an area of 30,000 square miles on the northwest side of the northeastern Appalachian Mountains belt, and extends into the adjacent Hudson, Champlain, and St. Lawrence Valleys to the northwest. It is athwart a major change in trend of this belt from northerly to northeasterly. The synthesis is a rationale of the tectonic relations of this region, discussed in chronological order.

The Precambrian basement, exposed at the core of a Paleozoic anticlinorium in the mountain belt, is made up of complexly deformed diaphthoritic miogeosynclinal rocks intruded by granitic plutons and pegmatite dikes. The exposed rocks are part of a northeast-trending Precambrian mobile shelf at least 300 miles wide that includes a wide belt to the northwest in the North American craton.

The lower (Cambrian and Ordovician) and middle (Silurian and Devonian) Paleozoic orthogeosyncline, which coincides mainly with the Appalachian belt, includes a broad eugeosynclinal zone, a miogeosynclinal zone to the northwest and probably also to the southeast of the eugeosynclinal zone, several geanticlines, and a quasi-cratonic belt; it thus contrasts with the broadly miogeosynclinal Precambrian rocks of the basement. The eugeosynclinal deposits, whose maximum thickness is more than 50,000 feet, average at least three times the thickness of those in the miogeosynclinal zone. Pelitic and semipelitic rocks dominate the upper deposits and lap over geanticlines, a quasi-cratonic belt, and the margin of the craton. The sources of sediments include cratonal areas, geanticlines that formed tectonic islands, and volcanic islands.

The transition between the miogeosynclinal and eugeosynclinal zones is one of sedimentary facies and thickness change, and of stratigraphic convergence and unconformity. In the lowest Paleozoic rocks the transition is west and northwest, respectively, of the Green and Sutton Mountains. The miogeosynclinal zone is missing in Quebec northwest of the northern Sutton Mountains and the eugeosynclinal zone extends to the northwestern margin of the orthogeosyncline. The belt of transition, however, moved southeast in younger rocks, so that in the middle Paleozoic rocks it lies between the Green, Sutton, and Notre Dame Mountains and the Connecticut and St. Johns Rivers.

Unconformities indicate stillstand in the miogeosynclinal zone, and general uplift of the northwestern part of the orthogeosyncline, followed by subaerial denudation of the geanticlines in both the eugeosynclinal and the miogeosynclinal zones and by repeated geosynclinal folding. The unconformities are within the lower Paleozoic (especially beneath the Middle Ordovician), are the most extensive between the lower and middle Paleozoic, and are within the middle Paleozoic (especially beneath the Lower Devonian).

Geanticlines, two of which coincide with gravity highs, are recognized by unconformable overlap and convergence of bedded rock units toward their axes. The lower Paleozoic Vermont-Quebec geanticline is northwest of the Green and Sutton Mountains in northwestern Vermont and neighboring parts of Quebec, but to the south and northeast it swings more into line with the mountains. It coincides with the lower Paleozoic belt of northwest-southeast transition from the miogeosynclinal to the eugeosynclinal zone, except near the north end of the Sutton Mountains where it trends into the eugeosynclinal zone. The lower Paleozoic Stoke Mountain geanticline coincides with the Stoke Mountains in Quebec, contains eugeosynclinal lower Paleozoic rocks, and is a little northwest of the belt of transition southeastward between the middle Paleozoic miogeosynclinal and eugeosynclinal zones. The lower and middle Paleozoic Somerset geanticline, which nearly coincides with the upper Connecticut River valley and the Boundary Mountains between Quebec and Maine, is cored by rocks of the lower Paleozoic eugeosynclinal zone and is truncated by the unconformity beneath middle Paleozoic rocks.

The distribution of the preorogenic igneous rocks of the eugeosynclinal zone reflects the southeastward retreat of this zone during the lower and middle Paleozoic. These rocks include mafic to intermediate metavolcanic and hypabyssal bodies, prevailingly of oceanic theoleiitic composition, and mafic and ultramafic plutons. The plutons reach Lower Cambrian to possibly Middle Ordovician stratigraphic levels.

The eugeosynclinal zone, and to a lesser extent the miogeosynclinal zone, are sites of regional metamorphism, shown most universally by the foliation that began to form with compaction of the shales. Isograds climb from low stratigraphic levels in the geanticlines to higher stratigraphic levels in the intervening geosynclinal troughs, showing a direct correlation between the thickness of bedded rocks and metamorphic intensity. Zones of highest grade metamorphism coincide with uplifts, but were probably originally deepest in the geosynclines. Undeformed garnet and staurolite-kyanite coincide with domes and arches, and deformed garnet and chloritoid-kyanite zones with anticlines. Some staurolite and sillimanite zones adjoin granitic plutons, but others are not so associated. Retrograde metamorphic effects in the Precambrian basement include replacement of garnet and hornblende by biotite and chlorite, and sillimanite by muscovite; these effects are caused by folding of the dry basement with the wet Paleozoic. In wet Paleozoic rocks, midway between the dry terranes of the basement and the domes and arches, garnet was replaced by chlorite and kyanite was replaced by muscovite as a result of uplift, denudation and cooling.

In Quebec, an exogeosyncline containing about 5000 feet of rocks overlies the northwestern part of the orthogeosyncline and the adjoining craton northwest of the Sutton Mountains. This is a secondary geosyncline northwest of the

Vermont-Quebec geanticline. It contains Upper Ordovician sandstone, shale, and limestone which overlie shale at the top of the lower Paleozoic miogeosynclinal zone but which are eroded from the eugeosynclinal zone.

The orthogeosyncline swings through the wide bend of the northwesterly bulge of the New England salient in northern New England and adjacent Quebec—all facies zones and tectonic features show a similar salient. It is deepened near the axis of the salient in a transverse trough that contains as much as 80,000 feet of strata. This section thins by stratigraphic convergence to less than 50,000 feet toward the flanks of the salient. The rocks are most varied in the salient, but thick sections of mafic volcanic rocks and carbonaceous pelites are characteristic.

Similar (passive and flexural flow) folds confined to the lower and middle Paleozoic bedded rocks and commonly overturned to the northwest toward the craton, are of an early regime of variously oriented folds. Cross folds, chiefly minor folds, trend northwest at right angles to the northeast structural trends. Longitudinal folds, slides, intrastratal intrusions, and syntectonic bodies of ultramafic rock that intruded in the solid state parallel the latter trends. Oblique folds parallel the flanks of the New England salient and swing into continuity with the longitudinal folds northeast and south of the salient and at its axis. The largest major longitudinal folds are thousands of feet above the Precambrian basement. The recumbent middle Paleozoic Skitchewaug nappe, the best known of the major longitudinal folds, is rooted to the southeast in the eugeosynclinal zone. Other recumbent folds exist, but their relations are more controversial. A middle Paleozoic intrastratal diapiric fold has been described west of the Skitchewaug nappe. Major longitudinal folds northwest of the Stoke Mountain geanticline underlie a Middle Ordovician unconformity; others in the same area are truncated by a pre-Silurian unconformity. Longitudinal folds on the Vermont-Quebec geanticline in the vicinity of the international boundary are nearly upright rather than overturned to the northwest. The lower Paleozoic Taconic slide is beneath Cambrian and Lower and Middle Ordovician eugeosynclinal rocks in the Taconic klippe and above autochthonous miogeosynclinal rocks of the same age west of the Vermont-Quebec geanticline and south of the New England salient. The oblique folds face southwest on the south flank of the New England salient in general harmony with the west-facing longitudinal folds, but on the northeast flank of the salient they face southeast. The largest of the oblique folds, like the large longitudinal folds, are thousands of feet above the basement.

The early folds and slides were produced by laminar flow and slip and by minimal flexing and thrusting, principally to the northwest. Several episodes of uplift in the eugeosynclinal deposits are probably accountable. The New England salient provided a basement framework that deflected, blocked, or reversed the northwestward movements to form especially the oblique folds and possibly the cross folds. The westward movement of the Taconic slide was probably assisted by maintenance of fluid pore pressure in the root zone near the top of the Vermont-Quebec geanticline by means of westward migration of water expelled from the thick eugeosynclinal deposits to the east during metamorphism.

The semiconcordant ultramafic, mafic, and intermediate intrusive rocks in the eugeosynclinal zone are subparallel to the foliation of the bedded rocks and syntectonic with the early longitudinal folds. The ultramafic rocks were emplaced in a solid and cool state, after transport that is interpreted as northwestward movement as the enclosing strata were folded. The less widely distributed gabbro and diorite, which lost their mobility with crystallization from magma, participated less actively in the folding.

Concordant calc-alkalic plutons, also emplaced in eugeosynclinal rocks, are synkinematic magmatic features that are less commonly parallel to foliation than are the semiconcordant intrusive rocks and are truncated upward by unconformities at successively higher levels in the direction of southeastward offlap of the eugeosynclinal zone.

Regional foliation, subparallel to both the axial surfaces and limbs and the axial-plane cleavage of the early folds, approaches parallelism with the bedding in most places inasmuch as early minor folds are sparse. Thus restored, the foliation conforms to the geosynclines and geanticlines, masking the Taconic slide. Sericitic mica and fine-grained chlorite, the principal foliate minerals, are features of low-grade regional metamorphism that progressed upward as the eugeosynclinal deposits accumulated, as shown by successive unconformities that mark sharp upward decreases in the foliate condition of the bedded rocks.

A longitudinal tract of middle Paleozoic domes and arches, characterized by drag folds that face downdip and some of which are cored by Precambrian basement rocks, trends northeastward across the Vermont-Quebec geanticline in southern Vermont. Largest in this tract is the Strafford-Willoughby arch which extends about equal distances northeast and southwest of the axis of the New England salient. The reverse drag folds are in the regional foliation, which near the crest of the domes and arches is obliterated by a new foliation that parallels the axial surfaces of the drag folds. The reverse drags indicate that the domes and arches were raised by vertical upward pressure, probably of buoyant rock beneath.

Grossly parallel (concentric) or flexural folds that trend northeast with the Appalachian structural trends, and the largest of which include the Precambrian basement, are of a late, middle Paleozoic regime. Smaller and variously oriented steeply plunging folds of this regime are above the basement. The principal form surfaces of these folds are the regional foliation in the eugeosynclinal zone and the bedding in the miogeosynclinal zone. Thrust faults, also of this regime, parallel the trend of the major folds. Axial-plane cleavage varies from fracture cleavage through crenulation cleavage to slip cleavage and slip-cleavage schistosity. The parallel fold style gives way to similar (passive-slip and flow) folds in parts of the eugeosynclinal zone. In these parts the form surfaces are offset on the axial-plane cleavage in directions both the same and the opposite of that of flexural drag folds, and the offsets opposite in sense predominate, accentuating the amplitude of the folds. Mineral lineations, less commonly slickensides, and some minor folds plunge downdip on the bedding and bedding foliation near thrust faults and on steep homoclinal limbs of major folds.

The late folds form anticlinoria that rudely coincide with the previously formed geanticlines, and synclinoria that coincide with the intervening and adjoining geosynclinal troughs. The axial surfaces of most folds in and south of the New England salient dip steeply southeast and the folds face northwest, but to the north of the axis of the salient the folds are nearly upright. The folds are also nearly upright in eastern Vermont, New Hampshire, and neighboring areas. The orientation of the axial surfaces of the folds changes gradually to subparallel with the flanks and tops of the domes and arches as the latter are approached. The folds in the northwestern part of the orthogeosyncline are tipped over to the northwest toward the craton, and the thrust faults in this same belt dip east in the same direction as the axial surfaces of the folds. The late folds of first magnitude are, from northwest to southeast, the Middlebury-Hinesburg-St. Albans synclinorium, the Green Mountain-Sutton Mountain anticlinorium, the Connecticut Valley-Gaspé synclinorium, the Bronson Hill-Boundary Mountain anticlinorium, and the Merrimack synclinorium.

The Middlebury-Hinesburg-St. Albans synclinorium is a major foreland fold in lower Paleozoic miogeosynclinal rocks and correlative allochthonous eugeosynclinal rocks of the Taconic klippe. This synclinorium is bordered to the west and east by thrust faults, which are most extensive on the south flank of the New England salient. The Green Mountain-Sutton Mountain anticlinorium, containing chiefly lower Paleozoic eugeosynclinal rocks, coincides with the Vermont-Quebec geanticline in the Green Mountains in central Vermont and the Notre Dame Mountains in Quebec, but near the axis of the New England salient it is southeast of the geanticline. The Connecticut Valley-Gaspé synclinorium, which contains middle Paleozoic rocks transitional from the miogeosynclinal to the eugeosynclinal zone, coincides with a geosynclinal trough between the Stoke Mountain and Somerset geanticlines and southeast of the southern part of the Vermont-Quebec geanticline. The configuration of the folds in the synclinorium is determined principally by the domes and arches near the synclinorial axis. The Bronson Hill-Boundary Mountain anticlinorium, which contains both lower and middle Paleozoic rocks, coincides in its northern parts with the Somerset geanticline. The Merrimack synclinorium to the southeast, which also contains lower and middle Paleozoic rocks, is a relic of a geosynclinal trough southeast of the Somerset geanticline.

The late folds and thrust faults were probably produced by subhorizontal movements as part of outward spread from the rising domes and arches. Rocks moved from the southeast into the New England salient. Steeply plunging minor folds, free from basement control, evolved in response to horizontal adjustments between major folds, and domes and arches in the thick eugeosynclinal section in the salient. Thrust faults evolved in the miogeosynclinal rocks south of the axis of the salient. The resulting counterclockwise movement of the thrust slices and their included folds and the folded rocks to the east of them in the eugeosynclinal zone continued until the present northward trend was achieved.

Monoclinal flexures and related kink layers, which dip northwest and parallel to which rock to the northwest was displaced upward and to the southeast, have been recognized in north-central and northwestern Vermont.

Joints include systematically oriented undeformed planar sets that dip almost vertically and cross the trend of the longitudinal folds and thrust faults at large angles. They also include less extensive nonsystematic joints that are curved or irregular and that end against the systematic joints. Some conjugate joint sets, the bisectrices of whose acute angles trend at right angles to the axes of the longitudinal folds, are possibly shear joints. Tension produced by bending of folds into the New England salient seems a doubtful cause of the joints, especially in the thick and deeply confined rocks of the eugeosynclinal zone.

Discordant and commonly nonfoliate middle Paleozoic calc-alkalic plutons are postkinematic magmatic features randomly emplaced in rocks deformed in both the early and late folds. They are most abundant near the axis of the New England salient where it crosses the eugeosynclinal zone.

Superimposed unconformably on the southeastern part of the orthogeosynclinal belt is an epieugeosyncline, containing upper Paleozoic clastic coal-bearing rocks. Before being eroded it probably covered wider areas of the eugeosynclinal zone, especially in the Merrimack synclinorium.

Systems of early Mesozoic high-angle faults, made up of nearly parallel longitudinal sets, strike north-northeast south of the axis of the New England salient and northeast north of the salient axis, parallel to the trends of the late longitudinal folds. Faults in the foreland belt of the Champlain-St. Lawrence Valley are downthrown to the southeast of domal structural features, and others in the Connecticut Valley are downthrown mainly to the northwest of similar features.

Lower Mesozoic terrestrial clastic and mafic volcanic (and hypabyssal) rocks unconformably overlie the eugeosynclinal zone and the epieugeosynclines (in taphrogeosynclines bounded by the high-angle faults) in southern New England and the Maritime Provinces.

Discordant and nonfoliate Mesozoic alkalic plutons are in curvilinear tracts that transect both the orthogeosynclinal belt and the craton. Dike rocks, also alkalic, are associated with the plutons and occur widely in areas between the plutons.

The chronology of the region is supported by biostratigraphic, radiometric, and structural data punctuated by unconformities.

The Precambrian chronologic record is inherently scanty. Metasedimentary basement rocks exposed in the Green Mountain-Sutton Mountain anticlinorium in Vermont provide late Precambrian radiometric ages and a regional metamorphic overprint dating from about a billion years ago. Comparable metasedimentary rocks in the basement of the Adirondack Mountains were deposited in the late Precambrian. Pegmatites provide radiometric ages about the same as those of the metamorphic overprint, which records erosional unloading and cooling that restarted the potassium-argon systems about 0.4 b.y. before the end of the Precambrian.

The earliest Paleozoic rocks, which are assigned to the Cambrian(?), overlie the Precambrian basement unconformably and are overlain conformably by miogeosynclinal strata containing fossils of Early, Middle, and Late Cambrian and Early and Middle Ordovician age. All epochs of the Cambrian and Ordovician are represented in the eugeosynclinal zone, and Late Ordovician fossils are found in the exogeosyncline. Potassium-argon radiometric values corresponding to Middle and Late Ordovician are mostly hybrids, between Cambrian to Early Ordovician metamorphic dates and the dates of widespread middle Paleozoic metamorphic overprints that with yet later overprints, have been revealed by Rb-Sr whole-rock isochron dating. Granitic plutons emplaced in Middle Ordovician rocks have yielded Middle or Late Ordovician Rb-Sr whole-rock isochron ages.

Quasi-cratonic middle Paleozoic strata, eroded from the Champlain-St. Lawrence Valley belt, were probably Upper Silurian or higher. Chiefly in the miogeosynclinal zone, or in comparable thin lithofacies, southeast of the Green Mountain-Sutton Mountain anticlinorium, are Early, Middle, and Late Silurian and Early and Middle Devonian fossils. The middle Paleozoic K-Ar values suggest mainly the time of metamorphism and several are probably hybrids of early dates and true Devonian dates. Southeast of the belt of rocks of hybrid ages is a belt that shows true K-Ar dates of Middle (?) Devonian Acadian metamorphism and deformation about 360 m.y. ago; this belt is without later (Appalachian?) metamorphic overprint, contains Early Devonian fossils, and its tightly folded strata are overlain unconformably by gently flexed strata of Middle Devonian age. Discordant calc-alkalic granitic plutons of comparable radiometric age transect some of the late folds. Rb-Sr whole-rock and Pb/alpha determinations to the southeast in the area of the post-Devonian overprint approximate the Acadian metamorphic date.

The late Paleozoic chronology in the northwestern New England and Quebec region is limited to a middle Permian metamorphic overprint with a K-Ar age of 250 ± 10 m.y. in the Merrimack synclinorium and environs. Unmetamorphosed felsic volcanic rocks that lie unconformably on the metamorphic rocks are possibly of Permian age. If this age is correct, the unconformity marks the Appalachian orogeny. The Mesozoic chronology is furnished by high-angle faults that bound the Late Triassic taphrogeosynclinal deposits in southern New England, and by alkalic intrusives of various radiometric ages (96 m.y.-to-180 m.y.), that intersect or are transected by the faults. The Cenozoic chronology is recorded by valleys and uplands produced by a continued selective downwasting, by Tertiary residual deposits containing lignite that were let down into valleys formed partly by solution of carbonate rocks, and by various Quaternary features related principally to glaciation.

The orthogeosyncline was formed in the earliest Paleozoic or possibly the latest Precambrian. The Vermont-Quebec geanticline started to form during the Cambrian by tectonic stillstand relative to subsiding adjacent geosynclinal troughs; other geanticlines probably first appeared in the Early to Middle Ordovician. As the geosynclinal troughs subsided, especially in the New England salient, volcanics were extruded and sediments that were derived from the craton, from geanticlines, from volcanic accumulations, and from intrageosynclinal uplifts, were deposited mainly in the troughs. Ultramafic, mafic, and intermediate plutonic rocks were first emplaced at the end of the Cambrian or beginning of the Ordovician in the eugeosynclinal zone, and the ultramafics were transported northwestward tectonically as serpentinization continued. Albitic granitic plutons were emplaced in the Early or Middle Ordovician. Regional foliation that had first appeared in the Cambrian as the geosynclinal troughs subsided continued to form. In the Middle Ordovician, stillstand of the Vermont-Quebec and Stoke Mountain geanticlines gave way to general uplift and denudation which included a westward sliding of the Taconic allochthon. During the late Middle and Late Ordovician, the Somerset geanticline appeared, granitic rocks were emplaced, and the other geanticlines continued as sources of sediments deposited in adjacent geosynclinal troughs, including the exogeosyncline. New generations of early folds formed as geosynclinal subsidence and uplift was renewed. The early Paleozoic closed with general uplift and erosion at and northwest of the Somerset geanticline, culminating in the climax of the Taconic disturbance.

The northwestern part of the orthogeosyncline stabilized to a quasi-cratonic belt early in the middle Paleozoic. The miogeosynclinal zone overlapped southeastward on the eugeosynclinal zone and eventually across the Stoke Mountain geanticline in the Late Silurian and Early Devonian. In the Late Silurian, rapid subsidence resumed in a geosynclinal trough southeast of both the Stoke Mountain geanticline and the southern part of the Vermont-Quebec geanticline that contained the belt of lateral transition from the miogeosynclinal to the eugeosynclinal zone. Meanwhile, the Somerset geanticline continued as a source of part of the eugeosynclinal clastics. An intrageosynclinal uplift from which sediments, recumbent folds, and intrastratal diapiric folds moved northwest and possibly southeast, probably formed in the geosynclinal trough southeast of the Somerset geanticline. Mafic to felsic intrusive rocks, especially concordant calc-alkalic plutons, continued to be emplaced and the regional foliation continued to form in the eugeosynclinal zone. Growth of the domes and arches and concomitant evolution of the late longitudinal folds and thrust faults during the Acadian orogeny climaxed the middle Paleozoic. The discordant calc-alkalic plutons were emplaced soon after, and then, 360 m.y. ago, northwest of the Merrimack synclinorium uplift and erosion followed, as did cooling, opening of joints, and restarting of K-Ar systems.

In the late Paleozoic the Merrimack synclinorium stood still, or possibly resumed subsidence to form the northwestern extremity of the epieugeosyncline that is preserved in southeastern New England. The epieugeosyncline was folded, faulted, and uplifted in the Appalachian orogeny, and then, with the adjacent Acadian uplift, was deeply eroded in the early Mesozoic. The taphrogeosyncline in southern New England was formed in the early Mesozoic and was followed in the middle Mesozoic by the alkalic intrusives. Thereafter until the present time, the Paleozoic and Mesozoic terranes were selectively weathered and eroded, and streams that possibly survived from the Appalachian orogeny flowed northwestward in northwestern New England and adjacent Quebec.

Introduction

LOCATION AND SCOPE

The exceedingly complicated tectonic relations of northwestern New England and adjacent Quebec are here studied in the perspective of a region of about 30,000 square miles in the northeastern extension of the Appalachian Mountains belt, in the northeasternmost United States and southeastern Canada (Fig. 1). This region is on the northwest side of the Appalachians athwart a major change in the trend of the mountains from northerly in southern New England to northeasterly in northern New England and southern Quebec. It covers all of the state of Vermont and adjoining parts of New York, including the Champlain Valley and upper Hudson Valley, western and northern New Hampshire, northwestern Maine, and Quebec, including most of the St. Lawrence Lowlands. Relations elsewhere are briefly discussed where a knowledge of them is believed useful in interpretation. Geologic maps of the states of Vermont (Doll, Cady, Thompson, and Billings, 1961), New York, (Fisher, Isachsen, Rickard, Broughton, and Offield, 1961), and New Hampshire (Billings, 1955), scale 1:250,000, and of Maine (Doyle and Hussey, 1967), scale 1:500,000, are now published. Modern geologic maps of areas in the southeastern part of Quebec are at a scale of 1:63,360 or larger.

The synthesis here presented is essentially a rationale of the tectonic relations of the region. Its original purpose was to clarify the regional setting of an area of about 1600 square miles in north-central Vermont in which detailed fieldwork had been completed, but which was difficult to interpret from the fieldwork alone. The synthesis has grown through field reconnaissance, oral and written discussion with colleagues, and study of the available geologic literature, both published and unpublished. It has brought to light many unforeseen relations pertinent not only to north-central Vermont but to a much larger region.

METHOD

The various tectonic features and related bodies of rock are first described and interpreted as nearly as possible in the order of their first appearance. This mode of presentation was selected so that such entities as geosynclines, geanticlines, early folds and slides, and concordant calc-alkalic plutons that were formed in an early tectonic regime would be clearly separated in discussion from

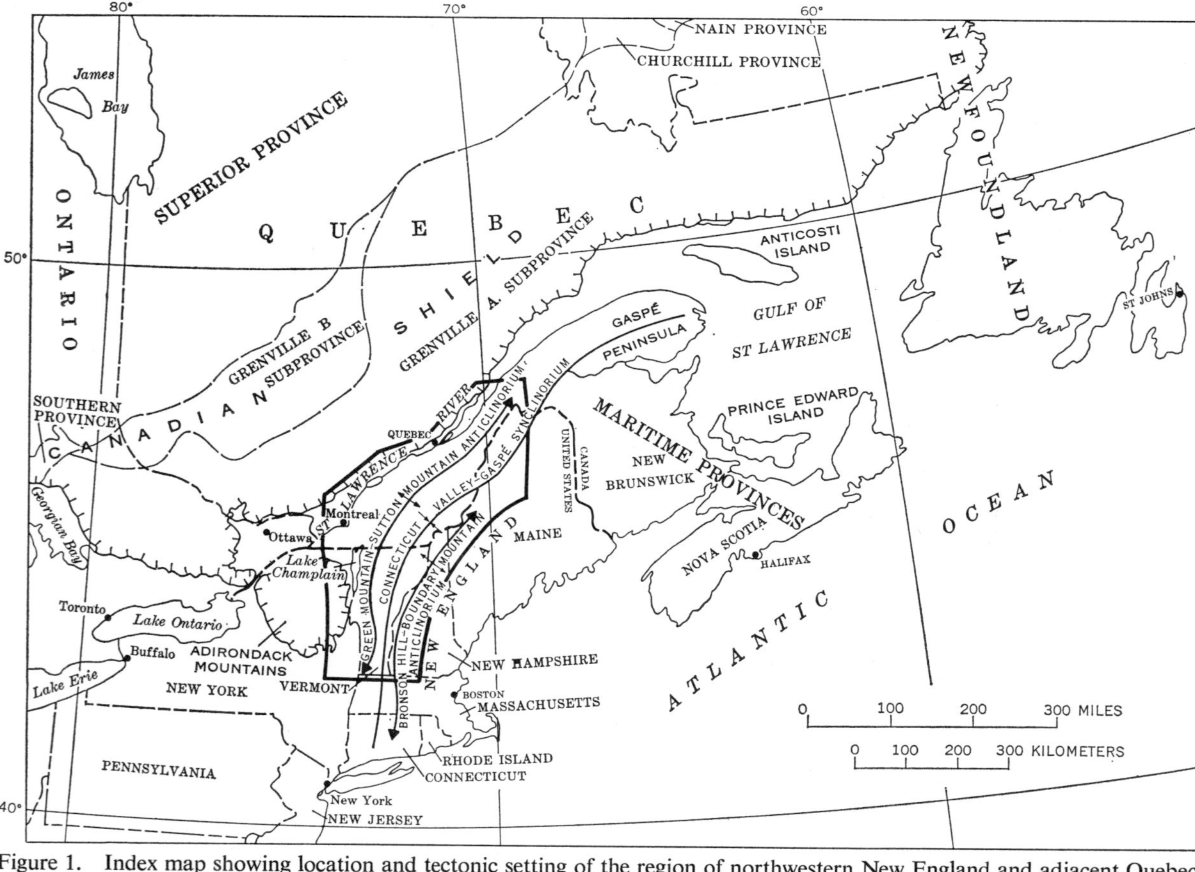

Figure 1. Index map showing location and tectonic setting of the region of northwestern New England and adjacent Quebec.

synclinoria, anticlinoria, late folds and thrust faults, and discordant calc-alkalic plutons that were formed during or soon after the late regime. The principal biostratigraphic and radiometric evidence for age of sedimentation, igneous intrusion, and metamorphism is then reviewed, and the discussion is closed with a running account of the chronology. Significant details that might interfere with the flow of discussion are included in footnotes or in the appendices. Access to discussions of sedimentary, magmatic, metamorphic, structural, stratigraphic, and geomorphic relationships, and of the related documents, which are considered under several headings in the varied light of the evolving tectonic setting outlined in the table of contents, is provided by the subject and author appendices.

The major tectonic features are shown on the regional tectonic map (Pl. 1). Many of the names preserve older ones or are familiar as revisions or extensions of names in present usage. The names of other tectonic entities, most notably geanticlines determined from stratigraphic studies but not clearly recognized from geologic maps, are recent or new. Geographic locations are specified chiefly with respect to the major tectonic features. Hence, the tectonic features listed in the major headings of the subject index provide locality information.

The rocks are discussed chiefly in their tectonic setting. Bedded rocks, for example, are discussed with reference to their eugeosynclinal or miogeosynclinal deposition, and intrusive rocks with reference to their relations to geosynclines and orogenies. The compositions and textures of the rocks are fairly reliable correlatives of tectonic setting. Stratigraphic details are omitted or discussed in footnotes, and the extremely involved history of stratigraphic research is reviewed elsewhere (Cady, 1960, p. 545–553, 571–576; 1968b). The major rock types and their structural features are summarized in the tectolithic sections (Pl. 2), in which the early folds and related slides of the geosynclinal regime are shown diagrammatically.

Basement Complex

The Precambrian basement complex crops out at the core of the southern part of the Green Mountain-Sutton Mountain anticlinorium and in adjacent domes to the east (Pl. 1). It is exposed widely a little to the northwest of the region studied, in the Canadian Shield and in the outlier of the shield that is in the Adirondack Mountains. The basement complex also crops out in various arched, domal, and upfaulted tracts to the south of the Green Mountain-Sutton Mountain anticlinorium, such as in the Berkshire and Housatonic Highlands in southwestern New England, and far to the northeast in western Newfoundland. Paleozoic rocks at the southeast side of the Appalachian Mountain belt are difficult to distinguish from (Quinn and Moore, 1968, p. 269–273) and widely cover Precambrian basement rocks, but the basement rocks are inferred in this part of the belt from occurrences in eastern Newfoundland and the Maritime Provinces. (*See* Isachsen, 1964, Fig. 1; Neale and others, 1961, Figs. 1 and 2; Poole, 1967, p. 17.)

The bedded rocks of the basement complex consist predominantly of gneiss and schist in which marble and quartzite units are interbedded. Amphibolite partly of mafic volcanic origin is known locally.[1] The igneous rocks include granitic plutons and widely distributed pegmatite dikes, but possibly most significant are the domical bodies of anorthosite and the apparently related gabbro, mangerite, and syenite, emplaced near the southeast edge of the Canadian Shield and in the Adirondack Mountains. (*See* Buddington, 1939, p. 208–209; de Waard and Walton, 1967, p. 605–608; Doll and others, 1961; Kranck, 1961, Fig. 1, p. 301, 306–314; Osborne and Morin, 1962, p. 119, 120; Stockwell, 1965b; Walton and de Waard, 1963.)

The basement rocks are complexly deformed, all are metamorphosed, and some are polymetamorphic. Precambrian and Paleozoic deformational and also metamorphic effects blend in the Paleozoic mobile belt southeast of the Canadian Shield and Adirondack area. In this belt, some basement units tend to swing parallel to unconformable contacts with overlying Paleozoic units and the basement rocks are diaphthoritic (Doll and others, 1961). In the Shield and Adirondack areas, on the other hand, the unconformities remain angular and the high-grade basement metamorphic rocks contrast with the unmetamorphosed Paleozoic strata that overlap them.

The exposed Precambrian bedded rocks are miogeosynclinal and presumably this is also true where they are buried by younger rocks. Marbles and quartzites extend southeastward from the Green Mountains for an unknown distance beneath Paleozoic cover. The marbles and quartzites (in the Grenville Series) trace northwestward in the Canadian Shield (through the "Grenville A Subprovince") to a line nearly parallel with and roughly 125 miles northwest of the St. Lawrence River. Northwest of this line (in the "Grenville B Subprovince") are eugeosynclinal rocks (Keewatin Series). These observations suggest a northeast-trending mobile shelf at least 300 miles wide. The northeastward and southwestward extent of the miogeosynclinal rocks is less clear. (*See* Appleyard, 1965, p. 55; Engel and Engel, 1953, p. 1042–1044; Engel, 1956; Grant, 1964; Grant, Wasserburg, and Albee, 1965; Osborne and Morin, 1962, p. 119–120, Fig. 1; Stockwell, 1965b; Wynne-Edwards, 1964, p. 54–56; Young, 1968, p. 117.)

The anorthosite and related rocks along the northwestern side of the suggested shelf area possibly also mark stable tectonic conditions during the formation of the Precambrian basement complex. Such bodies have been variously interpreted on the one hand as earmarks of water-poor, and tectonically, magmatically, and metamorphically consolidated terrane (Goldschmidt, 1922, p. 6–11; Osborne, 1937, p. 127) and on the other hand as evidence of wet environments without special reference to the tectonic and metamorphic setting (Buddington, 1960, p. 426–427, 429–430). The coincidence of the anorthosite bodies and the apparently miogeosynclinal environment seems significant at any rate, and indicates a need for more investigation and thought.

FOOTNOTES

[1] The bedded rocks of the Precambrian basement include the Mount Holly Complex at the core of the Green Mountain-Sutton Mountain anticlinorium in southern Vermont (Doll and others, 1961) and the Grenville Series of various authors in the Canadian Shield and in the Adirondack Mountains. The stratigraphic name *Grenville Series* refers properly to the type Grenville in Ontario at the southeast margin of the Canadian Shield. The Grenville Series is not to be confused with the term "Grenville province," or with rocks coextensive with the Grenville province whose metamorphism ceased with the cooling event that marked the end of the "Grenville orogeny." These rocks, which include the Grenville Series, are sometimes all referred to as "Grenville" or "Grenville Series," although they are of various parent facies that are both like and unlike and both correlate in time with and are older than the Grenville Series proper. These nomenclatural problems are touched on and various related interpretations of the Grenville problem are indicated in several recent publications by Appleyard (1965), Engel (1956; 1963, p. 150), Gilluly (1966, p. 102–109), Grant (1964), Osborne and Morin (1962), Stockwell (1961, p. 111–113, 116; 1962, p. 9, 13–14; 1963a, p. 126; 1963b, p. 126–128; 1964, p. 18–21; 1965a, 1965b; 1968, p. 695–697), and Wynne-Edwards (1964). (*See also* footnote 45, p. 116–117, and Appendix E.)

Orthogeosyncline[2]

GENERAL SETTING

The region is almost entirely within the northwestern border of the northern part of the lower and middle Paleozoic Appalachian orthogeosyncline, which coincides with the Appalachian Mountain belt and is adjoined on the northwest by the North American craton (Pl. 3; Kay, 1951, Pl. 1). The craton includes areas of Precambrian basement in the Adirondack Mountains and Canadian Shield and of thin Paleozoic cover in the intervening Ottawa Valley embayment. The orthogeosyncline (Cady, 1950b, p. 784) includes a broad and thick eugeosynclinal zone (primary geosyncline), a miogeosynclinal zone (mobile shelf)[3] to the northwest next to the craton, several geanticlines, and a quasi-cratonic belt superimposed on the northwestern part of the miogeosynclinal zone (Cady, 1960, p. 556, Pl. 2). Rocks transitional to miogeosynclinal type in southeastern New England suggest another miogeosynclinal zone, probably near the southeastern border of the orthogeosyncline.[4] The broad eugeosynclinal zone of the Paleozoic orthogeosyncline thus contrasts with the Precambrian basement complex whose tectonic affinities, as just noted, are largely miogeosynclinal. The orthogeosyncline was a mobile belt mostly of subsidence, locally and briefly of stillstand, and less briefly of general uplift.

EUGEOSYNCLINAL AND MIOGEOSYNCLINAL ZONES

Both zones contain variously metamorphosed lower and middle Paleozoic marine bedded rocks unconformably overlying the Precambrian basement. The eugeosynclinal zone is composed predominantly of slate, phyllite, and schist, with which are interbedded discontinuous and relatively thin metaquartzite, granofels, partly foliate granitic rock, calc-silicate rock, gneiss, greenstone, amphibolite, and metarhyolite units—metamorphic derivatives of interbedded shale and graywacke (collectively semipelites), minor quartz sandstone, calcareous siltstone, and mafic to felsic volcanic rocks. Some sandstone is quartz sandstone instead of graywacke, and where abundant suggests the miogeosynclinal zone. The volcanic rocks, especially the mafic, are the critical earmarks of the eugeosynclinal zone (Stille, 1940a, p. 8–9; 1940b, p. 14–15; 1950, p. 22–24). This zone is extrapolated into thick sections dominated by pelites and semipelites that although barren of volcanics are otherwise lithically and sequentially similar to sections that contain volcanics. The rocks of sedimentary origin contain large amounts of combined water, even in the metamorphosed condition. The miogeosynclinal zone includes quite continuous interbedded marble, quartzite, and slate units that are obviously metamorphic derivatives of the limestone, dolomite, quartz sandstone, siltstone and shale that remain unmetamorphosed near the margin of the craton. As a whole, the miogeosynclinal rocks are water-poor, especially where metamorphosed, and such combined water as they do contain is in the shale or slate. The miogeosynclinal rocks differ from those of the Precambrian basement in including relatively more quartz sandstone, dolomite, and limestone. Pelitic and semipelitic rocks, especially abundant in the upper part of the eugeosynclinal zone, overlap the margin of the craton, geanticlinal tracts, and the quasi-cratonic belt within the orthogeosyncline.[5]

Unmetamorphosed rocks are exposed west of the traces of the principal thrust faults, such as especially the Champlain thrust. Their unmetamorphosed condition must be made clear, inasmuch as they are potential sources of natural gas (L. R. Page, 1969, oral commun.; Taylor, 1964, p. 142, 144, 146; *compare with* Doll and others, 1961). They may extend several miles eastward beneath the metamorphosed hanging wall of the Champlain thrust in northwestern Vermont where displacement on the thrust is greatest (Cady, 1945, p. 577–578).

The bedded rocks in the eugeosynclinal zone average at least three times the thickness of those in the miogeosynclinal zone (Pl. 3). This thickness reflects greater subsidence and concomitant sedimentation and volcanic activity, and to some extent, repetition in recumbent folds and slides. The maximum thickness of the lower Paleozoic rocks in the miogeosynclinal zone is about 10,000 feet, whereas that of the thickest sections of rocks of the same age in the eugeosynclinal zone (Pl. 3, between sections *D-D′* and *E-E′*) is probably at least 50,000 feet. The source of the sediments in these great accumulations of eugeosynclinal rocks, especially the pelites and semipelites near their tops that overlap the geanticlines, is difficult to explain. Moreover, the geanticlines that formed tectonic islands within the orthogeosyncline do not appear to have been large enough to have provided the fairly abundant felsic sediments. However, volcanic islands are a sufficient source for at least mafic sediments. Possibly some, but certainly not all of the felsic sediments were derived from geanticlinal or cratonal sources to the southeast in the region now covered by the Atlantic Ocean (Boucot, 1968, p. 94; Naylor and Boucot, 1965). Continued areal geologic mapping and related stratigraphic studies will doubtless bring to light additional sources of sediments in southeastern New England (Appendix A).

The transition between the laterally coalescing miogeosynclinal and eugeosynclinical zones (Cady, 1960, p. 557; Osberg, in press) is marked by change in sedimentary facies (Logan, 1862, p. 323–325) and in total thickness of the rocks, and by thinning of stratigraphic sections (Pls. 1, 2, 3). This thinning toward the transition zone is chiefly by convergence and less commonly by unconformity. The transition in lowest Paleozoic rocks lies mostly west of the Green Mountains in Vermont and northwest of the Sutton Mountains in Quebec (Cady, 1960, Pl. 2; Doll and others, 1961). In Quebec the eugeosynclinal zone apparently extends to the northwest margin of the orthogeosyncline and no miogeosynclinal zone intervenes (Cady, 1960, p. 557–558). Lower Paleozoic rocks (Cambrian and Lower Ordovician) characteristic of the eugeosynclinal zone—graywacke, shale, and mafic to intermediate volcanic rocks—lie close to the Precambrian basement rocks of the craton and contain clasts locally derived from them (Melihercsik, 1954, p. 168; Osborne, 1956, p. 172–191, 197). The scarcity there of characteristic miogeosynclinal rocks has been attributed to overthrusting of the eugeosynclinal rocks upon them, but evidence for the implied extensive thrusting is equivocal (Clark, 1964c, p. 75–80; Osborne, 1956, p. 170–173), and it is difficult to explain how the coarse clasts derived from the craton crossed such a hypothetical miogeosynclinal zone. A less extreme counterpart of this anomalous distribution of the lower Paleozoic eugeosynclinal zone, up to the margin of the craton, is found in western Vermont. In this area some of the lowest Paleozoic (?), (Cambrian?) strata include mafic volcanic rocks and graywacke that lie conformably beneath rocks that are in and characteristic of the miogeosynclinal zone (Cady, 1960, Pl. 2, section *B-B′*; Pl. 3).

The transition from miogeosynclinal zone to eugeosynclinal zone moved southeast in the middle Paleozoic (Pl. 3) so that the northwest edge of the eugeosynclinal zone came to lie between the Green, Sutton, and Notre Dame Mountains and the Connecticut and St. Johns Rivers (Boudette and others, 1967, p. 28–45; Cady, 1960, p. 557, Pl. 2; Marleau, 1958a, p. 109; 1959, p. 137). Thin quartzose and calcareous units, whose affinities are miogeosynclinal although they are interbedded in the thick eugeosynclinal zone, extend southeast of the Connecticut River (Cady, 1960, p. 561–562). This southeastward overlap of the miogeosynclinal zone on the eugeosynclinal zone implies stabilization of the northwestern part of the orthogeosyncline and probable conversion of the lower Paleozoic miogeosynclinal zone to a quasi-cratonic belt (Cady, 1960, Pl. 2, p. 562).

UNCONFORMITIES

Unconformities and associated overlaps and stratigraphic convergences occur within the lower Paleozoic sections, both northwest and southeast of the Green and Sutton Mountains, and chiefly beneath Middle Ordovician strata (Albee, 1957; Cady, 1945, p. 537–539, 560; 1960, p. 564; Cady, Albee, and Chidester, 1963, p. 26; Lamarche, 1965, p. 60–61, 144–150, 1969, written commun.; Osberg, in press; Riordon, 1954, p. 8; 1957; *see also* Wanless and others, 1966, p. 79; St. Julien, 1963a, p. 30, 32, 125, 143–144, 193–195, 1965; Zen, 1964b, p. 28–30; 1967, p. 40–44; 1968, p. 134, 135–136). They mark movements that were forerunners of the climax of the Taconic disturbance[6], which in the northwestern New England-Quebec region is itself marked by a few angular unconformities and extensive disconformities (Pavlides and others, 1968, p. 64–75). These unconformities indicate episodes of stillstand mainly in the miogeosynclinal zone and general uplift of the northwestern part of the orthogeosyncline (the part of the orthogeosyncline in the region of the present study) that were followed by subaerial erosion and tectonic denudation of geanticlines in both the eugeosynclinal and miogeosynclinal zones, and by repeated folding within the geosynclinal deposits. Hiatuses at the unconformities within the lower

Paleozoic are rarely so great that formations are missing, except possibly in some southeastern belts of the Appalachians (Albee, 1961, p. C53; Boucot, 1961, Fig. 17, p. 157, Fig. 18, p. 158; Cady, 1960, p. 563; Green, 1968, p. 1606, 1629).

At the base of the middle Paleozoic in the miogeosynclinal zone, where that zone overlaps southeast on the eugeosynclinal zone southeast of the Green and Sutton Mountains, is the most extensive unconformity within the orthogeosynclinal rocks; it marks the close of general uplift and deformation in the Middle Ordovician-Early Silurian Taconic disturbance (Cady, 1960, p. 563; Clark, 1921), and in Quebec it marks an abrupt upward change from foliate to nonfoliate pelitic rocks. It extends for many hundreds of miles along the belt of overlap and also, in northeastern areas, as much as 100 miles southeast of the overlap into the lower and middle Paleozoic section of the eugeosynclinal zone (Green, 1968, p. 1606, 1629; Green and Guidotti, 1968, p. 262–264; Pavlides and Berry, 1966, p. B60; Pavlides and others, 1968, p. 64–75). Although the unconformity records a hiatus long enough to allow erosional unroofing of lower Paleozoic plutons (Cady, 1960, p. 563), the unconformity and the plutons signify only temporary interruption of the eugeosynclinal regime, for mafic volcanic rocks reappear in the middle Paleozoic strata southeast of the belt of overlap.

Unconformities occur within the middle Paleozoic section southeast of the Boundary Mountains, mainly beneath Lower Devonian strata (Boucot, 1961, p. 159–170).

Extensive angular unconformities are known only southeast of the region of the present study, where foliate lower and middle Paleozoic metamorphic rocks of the orthogeosyncline are overlain by less metamorphosed or unmetamorphosed rocks of secondary geosynclines (Cady, 1960, p. 563–564).

GEANTICLINES

Introduction

Several geanticlines paralleled the geosynclinal trend in northwestern New England and adjacent Quebec (Pl. 3) (Cady, 1967, p. 62–65) in the early and middle Paleozoic. They include tectonic islands where subaerial exposure is evident. Unconformable overlap and stratigraphic convergence of bedded rock units toward the axes of the geanticlines show their former existence. The geanticlines were relatively stable tracts of slowed deposition or nondeposition during subsidence of adjacent geosynclinal troughs and of local erosion during episodic general uplifts of the northwestern half of the orthogeosynclinal belt, and they were terranes of relatively mild deformation during subsequent orogenic folding. Two of the three known geanticlines coincide with present gravity highs, one of which at least was possibly caused by warping or faulting, or both, of density discontinuities in or at the base of the Earth's crust (Diment, 1956; 1968, p. 407–408). This possibility suggests that dense structural layers were left structurally high when adjoining geosynclinal troughs subsided. If the geosynclinal subsidence is an expression of crustal subsidence shown by gravity data, that data must reflect uncompensated anomalies that have persisted since the geosynclines first formed in early Paleozoic time—showing continued crustal strength from that time to the present (*see also* Innes and Argun-Weston, 1967, p. 74–76; Zen, 1967, p. 73). Whatever its basic cause, quasi-cratonic crustal strength evident in the vicinity of the geanticlines is to a degree comparable to that of the craton nearby to the west and is consistent with the miogeosynclinal affinities of the underlying Precambrian basement, from which some stability is probably inherited.

Despite their near coincidence with anticlinoria (to be discussed under a later heading), the geanticlines left as structural highs after subsidence of adjoining and intervening geosynclinal troughs are not to be confused with the anticlinoria produced later by folding under lateral compression. They probably tend to coincide because the early structural features guided the different later movement.

Vermont-Quebec Geanticline

The lower Paleozoic (Cambrian and Ordovician) Vermont-Quebec geanticline[7] is the most northwestern (Pl. 3). It is northwest of the axis of the northern Green Mountains and Sutton Mountains in northwestern Vermont and neighboring parts of Quebec but crosses the central Green Mountains in west-central Vermont and the northern extension of the Sutton Mountains in Quebec, then swings into rude coincidence with the Berkshire and Housatonic Highlands to the south in Massachusetts and Connecticut and with the Notre Dame Mountains to the northeast in more distant Quebec. It trends north along the lower Paleozoic belt of transition between miogeosynclinal and eugeosynclinal zones in New England and nearby Quebec, but near the north end of the Sutton Mountains in Quebec its northeastward trend puts it within the eugeosynclinal zone. It nearly coincides with a gravity high (Diment, 1953, Pl. 1; 1968, Fig. 30–3; Fitzpatrick, 1959, Pl. 2), whose strength in northernmost Vermont and adjacent Quebec is reinforced, at least in part, by dense mafic volcanic rocks (Tibbit Hill Volcanic Member of the Pinnacle Formation) exposed there (Diment, 1968, p. 403).

The Vermont-Quebec geanticline is marked in northwestern Vermont and immediately adjacent Quebec by unconformities in the eastern part of the miogeosynclinal zone (Cady, 1945, p. 537–538; 1960, p. 538–564), and by overlap of Middle Ordovician rocks upon Upper Cambrian ones (Cady, 1960, p. 541, 549, Pl. 2) near the transition into the eugeosynclinal zone. The stratigraphic section thins toward the axis of the geanticline (which here coincides with Enosburg Falls anticlinal tract shown on Pl. 1). The absence of 3000 feet of Lower Ordovician strata in northwestern Vermont reduces the miogeosynclinal section on the west flank of the geanticline to about 7000 feet, as compared with the 10,000 feet of Cambrian and Lower and Middle Ordovician strata farther west near northern Lake Champlain.

A little to the north in Quebec, the Vermont-Quebec geanticline is shown by tenfold stratigraphic convergence of Cambrian bedded rocks, including dolomite (Cady, 1960, p. 538; Clark, 1934, p. 10, Table 2), toward the geanticlinal axis.[8] The geanticline is also shown, where it trends northeastward from the miogeosynclinal into the eugeosynclinal zone (Pl. 3), by the intermittent lateral occurrence of quartz sandstone, dolomite, and limestone units in the Cambrian (Cady, 1960, p. 539, 558); the occurrence of these units suggests shallow water and stillstand relative to adjacent subsiding parts of the eugeosynclinal zone. The northeastern extension of this geanticline is perhaps the source of enormous blocks in limestone conglomerate and associated shallow-water clastic sediments that were spilled into the northwestern part of the eugeosynclinal zone in Quebec (Osborne, 1956, p. 197–198). Apparently similar accumulations, though of more local origin, are found west of the geanticline in the northern Lake Champlain region—New York, Quebec, and Vermont (Clark and Eakins, 1968, p. 167; Fisher, 1968, p. 30–32; Hawley, 1957, p. 68; Zen, 1968, p. 138).

Lower Paleozoic rocks have all been eroded in the central Green Mountains in west-central Vermont, but indirect evidence suggests that the Vermont-Quebec geanticline there coincided with the younger Green Mountain-Sutton Mountain anticlinorium (Pl. 3). Middle Ordovician strata, west of the Green Mountains, lap unconformably eastward over west-dipping older Ordovician and Cambrian units west of the axis of the anticlinorium (Cady, 1945, p. 560; Fowler, 1950, p. 35–37; Zen, 1964b, Pl. 2, sections *B-B′* and *C-C′*; 1967, p. 70). To the south, the section that lies unconformably beneath the Middle Ordovician west of the Green Mountains is downfaulted to the east (Thompson, 1959, p. 72), suggesting collapse of the west flank of the geanticline there during or shortly after the beginning of geosynclinal subsidence.[9]

The lower Paleozoic section at the north end of the Taconic Range, part of an allochthon (Cady, 1968a, p. 570; Zen, 1967) that moved westward from near the present site of the Green Mountain-Sutton Mountain anticlinorium (Hawkes, 1941, p. 661–663), also suggests the location of the Vermont-Quebec geanticline in the central Green Mountains (Pl. 3, section *E-E′*). The Cambrian and Ordovician section in the Taconic allochthon (4500 feet) in this area is thinner throughout than it is in either the eugeosynclinal zone east of the Green Mountains (20,000 feet) or the autochthonous miogeosynclinal section (7000 feet) to the west beneath the allochthon (Osberg, 1959, p. 45–46; Theokritoff, 1964; Thompson, 1952, p. 40; Zen, 1961, p. 297, 331–333). The rocks of the allochthon, though thinner, are lithically and sequentially most like the lower Paleozoic eugeosynclinal rocks exposed down the plunge of the Green Mountain-Sutton Mountain anticlinorium in northern Vermont and adjacent Quebec (Doll and others, 1963, p. 95; Zen, 1961, p. 333), near the axis of the geanticline.

The thinness of the section in the Taconic allochthon is probably explained by stratigraphic convergence and unconformable overlap of adjoining eugeosynclinal rocks to the east and miogeosynclinal rocks to the west, toward the Vermont-Quebec geanticline. This thinning, especially of the eugeosynclinal sand and shale clastics, has also been ascribed to the distance from their eastern source (Zen, 1967, p. 47; 1968, p. 131). On first consideration, this explanation would seem to make the geanticlinal interpretation of the thin section in the Taconic allochthon unnecessary. Westward attenuation of the eugeosynclinal clastics and similar eastward disappearance of miogeosynclinal clastics does not, however, seem to be the sole explanation of thinning, inasmuch as nonclastic carbonate rocks which predominate in the miogeosynclinal zone west of the original site of the allochthon also thin (eastward) toward the latter site. Had the geanticline been absent, carbonate rocks would probably have filled parts of the section at the source of the allochthon that were out of reach of the clastic sediments, thus maintaining a thickness at least equal to that of the miogeosynclinal zone. This seems clear because shallow-water carbonate rocks and clean quartz sandstones do actually fill in parts of the existing section in the allochthon (Zen, 1961, p. 303, 304, 306–307; 1964b, p. 16–17, 19, 21–22) as well as parts of its autochthonous equivalent exposed downplunge to the north (Clark, 1934, p. 10; 1936, p. 143, 146–149); conversely, dirty sand and shale-clastic rocks like those dominant in the allochthon and in northern autochthonous equivalents occur within the miogeosynclinal section (Osberg, 1959, p. 46; Shaw, 1958, p. 528–529, 536–537, 539–540, 547; Zen, 1961, p. 309–310; 1964b, p. 37–44). Thus the difference in water depth between the Taconic and miogeosynclinal settings was neither great nor irreversible (*see also* Zen, 1968, p. 131–133), or such as would alone cause thinning of either clastic or nonclastic sediments toward the Taconic section (*see also* Cady, 1968a, p. 570.)

Scarps, such as might mark the eastern edge of a carbonate shelf or bank, are probable (Rodgers, 1968). From such scarps some clastic carbonate material was distributed eastward (Bird and Theokritoff, 1967), and at some stratigraphic levels the scarps mark sharp facies contrasts (*see also* Zen, 1968, p. 131–132). However, in northwestern Vermont where their relations are most clear (Rodgers, 1968, p. 144), the sharp facies contrasts are at various intermittent stratigraphic levels, separated by typical miogeosynclinal carbonate rocks and quartz sandstones that extend farther east and thin out between the pelitic units where (approaching the geanticline) the sections converge stratigraphically and overlap unconformably eastward (Cady, 1960, p. 538–539; Dennis, 1964, p. 28–31; Stone and Dennis, 1964, p. 29–35, 40–43). The carbonate rocks in eastern as well as in western areas (most critically the Dunham, Bridgeman Hill, Rugg Brook, Saxe Brook, Rockledge and Gorge Formations) are sandy dolomites, calcareous quartz sandstones, and conglomerates that contain clasts of dolomite and limestone chiefly in a matrix of limestone and dolomite (*compare with* Rodgers, 1968, p. 144).

It is therefore probable that scarps of east-facing carbonate shelves or banks were neither distributed along a common line, nor necessarily continuous in time, and that where their record is best preserved they are west of the proposed Vermont-Quebec geanticline. (*Also* Cady, 1968a, p. 570–571.)

Tectonic thinning does not appear to have been involved in the relative thinness of the Taconic rocks, inasmuch as some of the least deformed sections (Theokritoff, 1964) are very thin. (*Also* Cady, 1968a, p. 571.)

Stoke Mountain Geanticline

The lower Paleozoic (Lower and Middle Ordovician) Stoke Mountain geanticline (Pl. 3) coincides with the Stoke Mountains of southern Quebec, which are southeast of the Sutton Mountains and trend northeast. It can be identified southward only to the Lowell Mountains, in Vermont. It contains eugeosynclinal lower Paleozoic rocks and is overlain by miogeosynclinal middle Paleozoic ones; the middle Paleozoic belt of transition from the miogeosynclinal to the eugeosynclinal zone is a little to the southeast. Upper Silurian and Lower Devonian miogeosynclinal rocks lap southeastward over the geanticline into the adjacent geosyncline. That the Stoke Mountain geanticline marks commencement of the consolidation of the orthogeosyncline with the craton (Cady, 1960, p. 556–557) is shown as follows: (1) A Middle Ordovician epieugeosyncline is proposed northwest of the Stoke Mountain geanticline in Quebec (Lamarche, 1965, p. 151; St. Julien, 1963a, p. 183–184), where the section appears to be nonvolcanic, although the sedimentary rocks resemble those interbedded with volcanics in the eugeosynclinal zone to the south in Vermont. (2) Clasts of albite granite derived from the geanticline are found in Middle Ordovician basal conglomerate (Lamarche, 1962; St. Julien, 1963a, p. 125). (3) Lower or Middle Ordovician basal conglomerate, exposed on the east side of the Lowell Mountains in Vermont (Albee, 1957; Cady and others, 1963, p. 25–27; Lamarche, 1965, p. 144–145) probably reflects the proximity of the southern extension of the Stoke Mountain geanticline. (4) The geanticline coincides with a poorly defined narrow belt of slightly higher gravity (Pl. 3) within a wider belt of southeastward-decreasing gravity southeast of the Vermont-Quebec geanticline (Diment, 1953, Pl. 1; 1968, Fig. 30–3).

Somerset Geanticline

The lower and middle Paleozoic Somerset geanticline[10] (Pl. 3) trends northeast near the upper Connecticut River and is about coincident with the Boundary Mountains along the frontier between Quebec and New Hampshire and Maine. At its core are bedded eugeosynclinal rocks that are Ordovician and perhaps also Cambrian in age (Green, 1968, p. 1607, 1610, 1613) and that have been variously assigned to the Cambrian or Ordovician, or both (Albee, 1961, p. C51-C52; Billings, 1956, p. 98), to a "basement complex" possibly Precambrian (Boucot, 1961, p. 184–185), to the Middle Ordovician (Cady, 1960, p. 554, 561, Pl. 3, annotation 21), and to the "pre-Upper Silurian" (Marleau, 1959, p. 133; 1968, p. 13). Any gravity highs connected with this geanticline are masked by gravity lows produced by low density metamorphic and plutonic granitic rocks. (*See* Joyner, 1963, Pl. 1.)

The unconformity that marks the Taconic disturbance and subsequent subaerial denudation truncates the Somerset geanticline and granitic plutons in the vicinity of the geanticline. The geanticline is overlain by Silurian quartz conglomerate, quartz sandstone, and limestone along its southeastern flank (Albee, 1961, p. 53–54; Boucot, 1961, Pl. 34; Cady, 1960, p. 563). This succession bespeaks stillstand and progressively lowered relief of the geanticline (Cady, 1960, p. 562), locally approaching the relatively stable condition of the miogeosynclinal zone. Elsewhere the Somerset geanticline appears to be overlapped unconformably by Silurian and Devonian eugeosynclinal rocks. A section of 50,000 + feet of Ordovician to Devonian eugeosynclinal rocks contained in and overlying the geanticline (Billings, 1956, p. 7, 9; Green, 1964, p. 11; 1968, p. 1606; Green and Guidotti, 1968, p. 255–262) may be compared with a section of possibly 65,000 feet of correlative eugeosynclinal and miogeosynclinal rocks west of the geanticline in north-central Vermont and with eugeosynclinal rocks of unknown thickness southeast of the geanticline in New Hampshire and Maine (Pl. 3). Lower Devonian eugeosynclinal rocks lap unconformably onto the Silurian units that are apparent near the northeast-plunging axis of the geanticline east of the northeast end of the Boundary Mountains (Boucot, 1961, Pl. 34).

MAGMATIC RELATIONS

The southeastward offlap of the eugeosynclinal zone is shown by both the horizontal and the vertical distribution of the initial magmatic rocks by which it is principally defined—mafic and intermediate(?) metavolcanic and hypabyssal rocks and variously metamorphosed mafic and ultramafic plutons tectonically of the preorogenic orthogeosyncline (Cady, 1960, p. 565). The mafic volcanic and hypabyssal rocks are predominantly oceanic tholeiites (Appendix B). The initial magmas were probably derived chiefly from oceanic crust beneath the eugeosynclinal zone (*see also* Diment, 1968, p. 407). Some of the initial magma, however, probably arose from lower parts of the continental crust beneath the northwestern part of the eugeosynclinal zone, with some contamination by the upper continental crust.

In lowest Cambrian(?) rocks of the Vermont-Quebec geanticline are volcanic greenstones (Tibbit Hill Volcanic Member of the Pinnacle Formation; Doll and others, 1961; *see also* Zen, 1967, p. 45–46). Higher in the section and southeast of the Vermont-Quebec geanticline are Lower Cambrian volcanic greenstone and amphibolite (Belvidere Mountain Amphibolite Member of the Hazens Notch Formation, and correlatives; Cady and others, 1963, p. 15–18; Doll and others, 1961). The southeastern limit of these volcanic rocks is indeterminate, inasmuch as Cambrian rocks are not exposed to the southeast within the region of the present study. Known most extensively in a northwest-southeast expanse in and south of the Stoke Mountain and Somerset geanticlines are Lower and Middle Ordovician volcanic rocks, chiefly greenstone, amphibolite, and metarhyolite, including quartz keratophyre (in the Stowe, Missisquoi, Ammonoosuc, Albee, and Dixville Formations in Vermont, New Hampshire, and Maine and in the Caldwell, Brompton, and Ascot Formations in Quebec). The metarhyolite is especially abundant in the Middle Ordovician rocks (Ammonoosuc Volcanics) found in and south of the Somerset geanticline. (Béland, 1957, p. 16–17; Billings, 1955, Cady and others, 1963, p. 24–25, 30–32; Doll and others, 1961; Green, 1968, p. 1608, 1612; Green and Guidotti, 1968, p. 257–259; Johansson, 1963, p. 27–28; Lamarche, 1967, p. 3–5; St. Julien, 1961b, p. 5–7; St. Julien and Lamarche, 1965, p. 4–6.)

The ultramafic rocks, predominantly serpentinite (Cady and others, 1963, p. 36–41, 57, 64; Chidester, 1953; 1962, p. 24–25), and the mafic plutons crop out principally in the northwestern part of the eugeosynclinal zone, southeast of the Vermont-Quebec geanticline. These intrusive rocks reach Lower Cambrian to possibly Middle Ordovician stratigraphic levels mainly in a belt along the southeast foot of the ridge of the Green, Sutton, and Notre Dame Mountains and are truncated by a Middle Ordovician unconformity in Quebec (Riordon, 1957; *see also* Lamarche, 1965, p. 153–154). Sediment derived from the ultramafites is found in Upper Ordovician strata (Lamarche, 1965, p. 155–157) that lie unconformably above them (Lamarche, 1962, p. 17; Osborne, written commun., 1964; St. Julien, 1963a, p. 227–229). Dunite, peridotite, pyroxenite, and gabbro are exposed mostly in that part of the belt northwest of the Stoke Mountain geanticline (Cady and others, 1963, Pl. 1; Cooke, 1937, p. 52–75; de Römer, 1963; Riordon, 1954, p. 8–12; St. Julien, 1961a, p. 7–8; 1961b, p. 10–11; 1965). The gabbros to the south in Vermont are in a much thicker section of bedded rocks, and are not widely exposed toward deeper parts of the geosyncline south of the Stoke Mountain geanticline. This may be because they are downdip and downplunge. Such a conclusion seems to be borne out by the reappearance of dunite in a relatively thin section near the west edge of the eugeosynclinal zone south of the New England salient in southern Vermont (Doll and others, 1961). Greenstone sills and dikes are emplaced in Ordovician strata and crop out mainly to the southeast of the ultramafic rocks in north-central Vermont and adjacent Quebec.

Thin Silurian volcanic greenstones and schistose, which are apparently rhyolitic volcanics (in the Shaw Mountain Formation), are exposed east of the Ordovician volcanic rocks in north-central Vermont, very near the upward transition from the southeastward-offlapping eugeosynclinal zone to the overlapping miogeosynclinal zone. Silurian(?) and Devonian volcanic greenstones and amphibolites are known on both flanks as well as south of the Somerset geanticline in eastern Vermont, western New Hampshire, northwestern Maine, and adjacent southeastern Quebec. (Standing Pond Volcanic Member of the Waits River Formation, volcanic rocks of the Littleton Formation, volcanic rocks of the Frontenac Formation—Billings, 1955; Cady and others, 1963, p. 33; Doll and others, 1961; Green, 1968, p. 1621; Green and Guidotti, 1968, p. 261; Marleau, 1958b, p. 5; 1959, p. 136; 1968, p. 23–25.)

Greenstone sills and dikes in eastern Vermont, western New Hampshire, Maine, and adjacent Quebec are emplaced in about the same Silurian, Devonian, and underlying strata as those that include the southeastern volcanic rocks (Billings and White, 1950; Cady, 1960, p. 565; Cady and Chidester, 1957; Green, 1968, p. 1623–1624; Hadley, 1950, p. 23–24). Ultramafic rocks and mafic plutons are known mainly in the northeastern part of the belt in Quebec and Maine where they intrude rocks believed to be as young as Devonian (Albee, 1961, p. C53; Green, 1968, p. 1622–1623; Green and Guidotti, 1968, p. 261; Lyons, 1955, p. 118; Marleau, 1959, p. 137–138; 1968, p. 34–37).[11]

The ultramafic intrusive rocks, the mafic and intermediate intrusive and volcanic rocks, and probably some of the felsic intrusive and volcanic rocks, are closely associated in the eugeosynclinal setting; the proportion of ultramafic rock is estimated at about one-tenth of the whole. (*See* Cooke, 1951; Doll and others, 1961.) This association suggests their mutual origin by fractionation of a complex magma that produced crystal accumulates—principally dunite and peridotite—and a residual magma from which chiefly basalt, gabbro, and diorite crystallized. The residual magma first produced volcanics and mafic to felsic intrusive rocks but was soon immobilized even at depth by crystallization. Thus magmatic intrusive rocks were finally emplaced near the site of fractionation, whereas the ultramafic intrusive rocks, though solid from the time of crystal accumulation, were mobilized with serpentinization, squeezed various distances from the site of fractionation, and cut some of the mafic intrusives (Cady and others, 1963, p. 45–46; Chidester, 1962, p. 88–89; 1968, p. 349–350; Chidester and others, 1951, p. 7–8; de Römer, 1963; Jahns, 1967, p. 143–144, 155–156; Riordon, 1954, p. 12).

METAMORPHIC RELATIONS

General Statement

The eugeosynclinal zone and to a much lesser extent the miogeosynclinal zone, are sites of regional metamorphism. This metamorphism kept pace with the deposition of the bedded rocks, beginning in the Cambrian(?) and Early Cambrian, and continuing on into later episodes of deformation and igneous intrusion. The effects of the metamorphism, especially foliation that began to form at low metamorphic grades and even during compaction of shales (Moench, 1966, p. 1452–1453; *see also* Maxwell, 1962, p. 294, 302), vary and depend directly upon several closely interrelated factors. These factors are: the amounts of pore water and clay minerals in the sediments, the load of overlying rocks, and the amount of deformation and therefore the active stresses within the rock. The varied intensity of metamorphism, as shown by the index minerals in the pelitic rocks, which range from chlorite to kyanite and sillimanite, is a function of pressure, temperature, and activity of water (Thompson, 1955, p. 98). The values of each of the latter variables were probably greatest in the eugeosynclinal zone. The abundance of porphyroblasts and mineral segregations in the eugeosynclinal zone also directly reflect increasing metamorphic grade, although not as precisely. Prograde metamorphism prevailed until very late in the Paleozoic rocks in the region, but retrograde effects are outstanding features of the Precambrian basement. Temperature and chemical activity, not only of water but of carbon dioxide and oxygen, are significant in the interpretation of tectonic relations. (*See also* Zen, 1963c, p. 930–936.) Generally speaking, both rock pressure and temperature increased with depth in the Paleozoic section, but where dry rocks in the basement are approached, the activity of water in younger units is inferred to have decreased nearing the basement contact (Appendix C).

Relation of Metamorphism to Thickness of Overlying Rocks

Isograds tend to rise from low stratigraphic levels in the thin sections over the geanticlines to higher stratigraphic levels in the thicker sections in the intervening geosynclinal troughs (Pl. 3). This is especially clear in the northwestern part of the orthogeosyncline, which is nearly free of the superimposed regional metamorphic overprints and the thermal metamorphic halos that surround plutonic igneous rocks. This rise of isograds to high stratigraphic levels in thick sections illustrates the direct correlation between thickness of the bedded rocks and intensity of prograde metamorphism—the troughs subsided more and consequently received greater thickness of sediments between any two stratigraphic levels. The increase in metamorphic intensity with thickness of overlying rocks is shown in the vicinity of the Vermont-Quebec geanticline in central Vermont, where lowest Paleozoic (Cambrian?) biotite-bearing stratigraphic units on the geanticline thicken into a geosynclinal trough and the index minerals change from biotite to garnet and chloritoid-kyanite. This change in metamorphic intensity is actually observed where the lowest Paleozoic rocks are exposed in the axial anticlines of the Green Mountain-Sutton Mountain anticlinorium in the central Green Mountains (Doll and others, 1961). In that area the axial anticlines trend northward from the Vermont-Quebec geanticline (at transition from miogeosynclinal to eugeosynclinal zone, as illustrated on Pl. 1) diagonally into the geosynclinal trough east of the geanticline. A similar change, although showing less contrast in metamorphic intensity, may be observed along a section at right angles both to the Vermont-Quebec geanticline (which here coincides with the Enosburg Falls anticlinal tract on Pl. 1) and to one of the axial anticlines of the Green Mountain-Sutton Mountain anticlinorium, in and west of the Sutton Mountains in Quebec.

The amount of deformation also increases in the direction of the thicker sections, where great internal mobility combined its effects with those of metamorphism connected with deep burial and related high temperatures and lithostatic pressures. Deformation and metamorphic intensity have been cited as, respectively, cause and effect in the Sutton Mountain area. In that area porphyroblastic albite and less commonly garnet and recrystallized mica decrease westward from the axis of the axial anticline of the Green Mountain-Sutton Mountain anticlinorium and are therefore considered peculiarities of the core of the anticline; moreover, some of the mica is in and parallel to crenulation cleavage that is in turn rudely parallel to the axial surface of the anticline (Rickard, 1964a, p. 90; 1965, p. 527). However, the albite porphyroblasts both predate and postdate the crenulation cleavage (*see also* de Römer, 1961, p. 279), and only those that predate the cleavage can be attributed to the deformation connected with formation of the anticline. The eastward increase in the intensities of deformation and metamorphism in this area seem quite as readily explained by the increase in thickness of the stratigraphic section eastward from the Vermont-Quebec geanticline; this thickening possibly accounts for mutually independent increases in deformational and metamorphic intensity. Quite clearly, albite, garnet, and mica continued to form during and after the anticline and its crenulation cleavage evolved. Their diminution westward, however, in a syncline on the east flank of the Vermont-Quebec geanticline is probably attributable to the much thinner section in and west of the syncline.

The Somerset geanticline, particularly in northeastern Vermont and northwestern New Hampshire, also exposes rocks of lower metamorphic grade than those in adjoining geosynclinal tracts. Lower and middle Paleozoic rocks of chlorite and biotite grade are adjoined on both flanks of the Somerset geanticline by rocks that average somewhat higher stratigraphically but contain garnet, kyanite, and staurolite. (*See also* Billings, 1955; Doll and others, 1961.) The effect of varying relative thickness of overlying rocks on intensity of metamorphism is here obscured by postgeosynclinal domes and arches to the northwest, and by concordant (synkinematic) granitic plutons to the southeast of the geanticline. Steeper than "normal" thermal gradients are associated with these features.

Relation of Metamorphism to Structural Features

The isograd patterns conform in varying degree with tectonic highs (*see also* Albee, 1968, p. 329–330). The highest grade rocks are at their centers and generally indicate increase of metamorphic intensity with depth in the geosyncline before they were formed.

Coincidence of the garnet and staurolite-kyanite zones and of commonly undeformed staurolite and kyanite porphyroblasts with domes and arches (Woodland, 1965, p. 133; oral commun., 1966) may be variously interpreted. This coincidence may be due to the somewhat delayed rise of geoisotherms with the same general configuration as the domes and arches—the latter grew more rapidly than the outflow of heat from beneath them, hence the "normal" (less steep) thermal gradient was not re-established until after the growth had stopped; or they may have resulted from the rise of heat from discordant (postkinematic) granitic plutons that crop out in the vicinity (Woodland, 1963, p. 363). Inasmuch as the plutons are in a belt that crosses the trend of the domes and arches and their associated metamorphic highs, it seems that rise of geoisotherms with the configuration of the domes and arches is more likely. A thermal gradient was first established before deformation, because some porphyroblasts of lower grade minerals, such as biotite and garnet, are deformed and their isograds have also been disturbed.[12]

The garnet and chloritoid-kyanite metamorphic zones coincide with the cores of anticlines, a relationship that may be interpreted in various ways, such as: (1) thermal anticlines, like those of domes and arches, evolved concomitantly with the structural anticlines; (2) the thick overlying rocks in the geosynclinal pile, affecting both temperatures and pressures, were responsible for the generally higher metamorphic intensities at depths that were revealed after folding and erosion, but only at the cores of anticlines; and (3) the dry quartz-feldspar rocks and rocks of sillimanite grade (before diaphthoresis) in the Precambrian basement provided a "sink" into which water contained in the lower Paleozoic bedded rocks that were undergoing metamorphism was able to flow. This reduced the partial pressure of water in the bedded rocks and thereby contributed to the general prograde metamorphic intensity revealed near the anticlinal cores. (*See also* Thompson, 1955, p. 99.)

Under all three of these suggested interpretations of the metamorphic relations of anticlines, prograde metamorphic intensity was regionally greater at depth before folding. In anticlines where porphyroblasts (including those of albite) that were formed during advancing metamorphism are deformed by the folds and deflect or are cut by axial-plane cleavage (Brace, 1953, p. 99; de Römer, 1961, p. 278–279; Osberg, 1952, p. 93–94, 99–100), an apparently greater metamorphic intensity resulted from upfolding of rocks of advanced grade. Where the growth of the porphyroblasts has continued after the close of folding (Rickard, 1964a, p. 90; 1965, p. 527), the thermal anticline may be an explanation of the greater metamorphic intensity. The desiccating effect of the Precambrian basement seems a probable factor in greater metamorphism of western anticlines of the Green Mountain-Sutton Mountain anticlinorium, where involved bedded rock units lie directly upon the basement. Thus the anticlines of the late longitudinal fold regime also mark folded isograds and geoisotherms.

Metamorphic granitic rocks including granite, quartz monzonite, granodiorite, and quartz diorite are exposed in domal anticlines transected by the biotite, garnet, staurolite, and sillimanite isograds.[13] They are in approximately the same stratigraphic position as metarhyolitic volcanic rocks (Ammonoosuc Volcanics) and as granofels (quartz-plagioclase granulite) and feldspathic quartzites, and less commonly phyllites (Middle Ordovician—Moretown Member of the Missisquoi Formation, Albee Formation, and Orfordville Formation) to the west and northwest in eastern Vermont and western New Hampshire (Billings, 1955; Doll and others, 1961). Homogeneity and primary foliations indicate a magmatic origin for some granitic rocks (Billings, 1956, p. 146–147; Page, 1937, p. 62–72). Those rocks considered to be metamorphic show regional concordance of their upper contacts with the bedded rocks and chemical similarity to known bedded rocks at about the same position; moreover they are cut by some calc-alkalic granitic rocks that are magmatic (Naylor, 1967, p. 18–28, 58–70; 1968, p. 233, 238–239; 1969, p. 410–412, 417–420). The metamorphic granites very likely extend beneath saddles and synclinal areas between domal anticlines in which they are exposed rather than being confined simply to the anticlinal cores. If so, they form, at least in part, a granitized stratigraphic unit transformed mainly from the felsic volcanic rocks. If not, the domal anticlines may be sites of thickest volcanic accumulation, before metamorphism (Naylor, 1968, p. 238; Thompson and others, 1968, p. 216–217). To the south in Massachusetts and Connecticut where the sillimanite zone is more widespread, the mafic volcanics (Ammonoosuc Volcanics) that overlie the granitic rocks change from amphibolite to quartz-feldspar and hornblende gneiss (Hadley, 1949; Lundgren, 1962, p. 7, 15); the gneiss was originally mapped as an intrusive igneous rock (Dana Diorite—Emerson, 1917, Pl. 10, p. 244–247).

The staurolite and sillimanite metamorphic zones do not seem as closely related to structural features as do the garnet, staurolite-kyanite, and chloritoid-kyanite zones; instead they seem related to granitic plutons (Bethlehem Gneiss and Kinsman Quartz Monzonite—Billings, 1955; 1956, p. 55–61). These plutonic sources of thermal energy are especially abundant on the northwest limb of a broad, probably shallow synclinorium (Merrimack synclinorium) in the interior part of the eugeosynclinal zone, southeast and south of the Somerset geanticline. Large areas of the sillimanite zone in this synclinorium are, however, devoid of intrusive rocks, at least at the ground surface (Billings, 1955), which makes it difficult to believe that magmatic heat was the sole cause of metamorphism, unless it is assumed that the igneous rocks are possibly at depth, or eroded (Hamilton and Myers, 1967, p. C15). Moreover, recent studies (Thompson and others, 1968, p. 215–216; Thompson and Norton, 1968, p. 325) show that the staurolite and sillimanite zones coincide closely with the axial traces of nappes that apparently brought their heat along with them from the interior of the eugeosyncline. (*See* Appendix A.)

Retrograde Metamorphic Relations

The retrograde metamorphic effects in the Precambrian basement rocks of the Paleozoic mobile belt are tied to simultaneous folding of these dry, miogeosynclinal, and already high-grade metamorphic rocks, and of the wet, low-grade rocks of the unconformably overlying Paleozoic. The basement probably once contained rocks of the sillimanite zone or higher grade, but now contains garnet and hornblende widely replaced by biotite and chlorite; moreover, incompetent basement rocks near the unconformity are foliated. Sillimanite in the basement rocks has apparently all been replaced by mica (Brace, 1953, p. 60–64; 1958; Doll and others, 1961). The retrograde effects were evidently dependent on water that probably flowed from the Paleozoic strata during metamorphism. (*See* Thompson, 1955, p. 99.) Paleozoic retrograde effects are unknown in the Precambrian of the Adirondack Mountains. There, unmetamorphosed Paleozoic strata form only a thin cratonal cover, Paleozoic deformation probably included only broad warping, and the Precambrian metamorphism of the "billion-year event" had been quenched under "dry" conditions—consequently without diaphthoresis—during subsequent uplift and denudation.

In the lower Paleozoic (Lower Ordovician Stowe Formation) in north-central Vermont, garnet was replaced by chlorite, and kyanite by sericitic muscovite. These retrograde metamorphic effects are probably an outcome of uplift and subsequent denudation and cooling of wet, eugeosynclinal, low- to middle-grade rocks and were realized only where enough water for the necessary hydration was still available in the geosynclinal pile. The diaphthoritic rocks crop out about midway between the Vermont-Quebec geanticline on the west that contains the known Precambrian basement, and post-geosynclinal domes and arches on the east, which involve full sections of lower and middle Paleozoic rocks (Albee, 1957; Cady, 1956; Cady, Albee, and Murphy, 1962; Doll and others, 1961). Probably the intermediate geographic position of these rocks left them the wettest in the geosynclinal pile after prograde metamorphism; thus water was available to form chlorite and sericite pseudomorphs. Plutonic rocks are an unlikely source of the water inasmuch as none, other than ultramafites, crop out in the diaphthoritic terrane. Prograde metamorphic terranes, both adjoining the known Precambrian basement to the west and centered on the domes and arches, are an unlikely source inasmuch as they are geographically remote from the diaphthoritic terrane. Instead the water would appear to have been produced by prograde metamorphism that continued at depth at the same time that the diaphthoritic terrane was being unloaded and cooled. The quartz-muscovite-garnet-kyanite schist that crops out in this terrane conceivably yielded water at depth, leaving sillimanite gneiss not yet exposed by erosion. (*Compare with* Albee, 1968, p. 329, 331.)

FOOTNOTES

[2] This discussion of the Appalachian orthogeosyncline in northwestern New England and adjacent Quebec appears in an earlier publication (Cady, 1967) in somewhat abbreviated form and without treatment of magmatic and metamorphic relationships.

The term *orthogeosyncline* refers collectively to troughs of primary geosynclinal subsidence and contiguous mobile shelves in a stage of development preceding orogeny—as opposed to *parageosynclines* (secondary geosynclines), formed during or after episodes of orogeny (Cady, 1950b; Stille, 1940b, p. 656–657). Together the orthogeosyncline and various parageosynclines compose the Appalachian mobile belt.

[3] The three-dimensional terms *eugeosynclinal zone* and *miogeosynclinal zone* emphasize the fact that the eugeosynclinal and miogeosynclinal rocks are commonly not the contents of discrete troughs—eugeosynclines and miogeosynclines—but merge laterally with each other within the orthogeosyncline at facies transitions whose geographic position may change with time (Cady, 1960, p. 557–559; 1967, p. 58–61; 1968b, p. 153; *see also* Kay, 1951, p. 67; Stille, 1940b, p. 653–654, 656).

[4] A miogeosynclinal zone has not hitherto been specifically recognized southeast of the eugeosynclinal zone in New England, but Precambrian(?) and Cambrian quartzite and limestone interbedded with schist and slate in Massachusetts and Rhode Island (Emerson, 1917, p. 24–31, 35–39) suggest rocks that are transitional southeastward from the eugeosynclinal zone to a miogeosynclinal zone.

[5] Most notable of the pelites and semipelites in the upper part of the section are those of Middle Ordovician age in the northwestern part of the orthogeosyncline. At the northwest near the craton along Lake Champlain, are the Cumberland Head, Stony Point, and Iberville Formations. Next southeast are the Hortonville, Morses Line, and Stanbridge Formations in the eastern Champlain Valley. Originally southeast of the latter formations, but now in the Taconic klippe, is the Pawlet Formation. Southeast of the Green and Sutton Mountains are the Cram Hill Member of the Missisquoi Formation and Magog Formation. Chiefly southeast of the Connecticut River are the Partridge and Dixville Formations. Other pelites and semipelites that are more to the southeast in the orthogeosyncline are in the Silurian Northfield Formation east of the Green Mountains, in the Devonian Gile Mountain Formation west of the Connecticut River, and in at least the type Devonian and Silurian(?) Littleton Formation (Billings, 1937, p. 487–490) southeast of the Connecticut River. (*See* Cady, 1960, Pl. 3; Billings, 1955; Doll and others, 1961; Green, 1964, Pl. 1; 1968, Pl. 1; Green and Guidotti, 1968, Fig. 19–2; Hatch, 1963, Pl. 1.) Many of these rocks are carbonaceous and contain syngenetic iron-bearing sulfides that give rise to characteristic rusty-weathering outcrops. Of especial tectonic significance is the fact that these rocks form the principal recumbent folds and the soles of most of the slides, discussed in a later section.

[6] The author (Cady, 1960, p. 564–565) has previously expressed doubt about a "pre-Taconic" unconformity southeast of the Sutton Mountains, inasmuch as at several places where this had been reported (Cooke, 1950, p. 34, 40, 42, 57–61, 76–79, 84–86, 88–91) they were not to be found. Nevertheless, the pre-Normanskill unconformity recently discovered by St. Julien (1963a, p. 125, 193–195) in the area between Magog and Sherbrooke, Quebec, and since confirmed by Lamarche (1969, written commun.) in the Disraeli area clearly shows there was at least one that predates the climax of the Taconic disturbance.

[7] The designation "Vermont-Quebec geanticline" takes the place of "Quebec Barrier," originally proposed by Ulrich and Schuchert (1902, p. 639) as a continuous linear subaerial topographic feature believed to have separated the eugeosynclinal zone from the miogeosynclinal zone. More recent studies indicate that such a complete separation of eugeosynclinal and miogeosynclinal zones does not exist in Vermont and Quebec (Cady, 1945, p. 561; 1960, p. 557). The Vermont-Quebec geanticline appears to be homologous with the "ancestral Blue Ridge undation" of Bloomer and Werner (1955, p. 599, 601) in Virginia. The term "Vermont-Quebec geanticline" is not synonymous with "Quebec Geanticline" of Poole (1967, p. 25–26).

[8] Thinning of the Cambrian rocks on the Vermont-Quebec geanticline in Quebec, may be in part tectonic (*see* Eakins, 1964, p. 1; *also per* M. J. Rickard, written commun., 1965), but the spacing of bedding, notably in the dolomitic units, seems too little reduced compared with that of the thick sections west of the geanticlinal axis to indicate tenfold tectonic thinning (*see also* Clark and Eakins, 1968, p. 171).

[9] Zen (1967, p. 70–71, Fig. 15; 1968, p. 133–135) interprets the unconformity west of the Green Mountains in west-central Vermont differently, specifying that flexural folds preceded the unconformity and implying that longitudinal high-angle faults downthrown principally to

the west were nearly synchronous with it. He suggests that the faults are part of a "collapse structure . . . along the axis of the future Middlebury synclinorium." This interpretation seems very unlikely (Cady, 1968a, footnote 1, p. 566).

[10] Boucot, Field, and others (1964, p. 76), have identified a local source in Somerset County, western Maine, for material in Silurian and Lower Devonian sedimentary rocks. They refer to this source as paleogeographic "Somerset Island." The term geanticline seems more generally appropriate to include areas in which broad and continuous subaerial extent is less evident.

[11] Harwood (1967, Fig. 2) tentatively extends the Middle Ordovician Dixville Formation into lithically comparable unfossiliferous terrane in which ultramafites are emplaced and which has hitherto been mapped as the Devonian Frontenac Formation. This he does (written commun., 1967) partly because these are the only reported occurrences of middle Paleozoic ultramafites in the Appalachian Mountain belt and hence seem questionable.

[12] B. G. Woodland (oral commun., 1966) reported that biotite and garnet that had formed under static metamorphic conditions are deformed in an area between the Strafford dome and the Green Mountain-Sutton Mountain anticlinorium in east-central Vermont; staurolite and less clearly kyanite in this area fail to show the deformation. This condition contrasts with the situation in southeastern Vermont where garnet, kyanite, and staurolite formed under continuing dynamic conditions (Rosenfeld, 1960, 1965, 1968).

[13] The granitic rocks of the domal anticlines have been referred to the Oliverian Plutonic Series (Billings, 1955; 1956, p. 48–53; Page, 1968, p. 374–377). Syenite and hornblende quartz monzonite have also been included (Chapman, Billings, and Chapman, 1944, p. 514–515), but because of their structural and compositional affinities with the White Mountain Plutonic Series (Billings, 1955; Billings and Keevil, 1946, Fig. 3, p. 806) they are excluded here.

Exogeosyncline[14]

Superimposed on the orthogeosyncline and on adjacent parts of the craton northwest of the Green and Sutton Mountains is an exogeosyncline (Pls. 1, 3; Kay, 1951, p. 17–20). This is a secondary geosyncline that contains about 5000 feet of uppermost lower Paleozoic (Upper Ordovician) sandstone, shale, and limestone, preserved in southeastern Quebec where they conformably overlie shales mostly at the top of the Cambrian and Lower and Middle Ordovician miogeosynclinal zone and extend onto the craton (Clark, 1947, p. 6–15; 1955, p. 19–36; 1964a, p. 5–19; 1964b, p. 32–41; 1964c, p. 8–62). Near the margin of the craton in the vicinity of the city of Quebec, however, the Upper Ordovician rocks lie unconformably on Middle Ordovician rocks of the eugeosynclinal zone (F. F. Osborne, written commun., 1967). The southeastern extent of the exogeosyncline is undetermined, inasmuch as its rocks are eroded from all but the St. Lawrence Valley area. However, the Vermont-Quebec geanticline, which supplied the clastic sediments, is the likely southeastern limit.

[14] Nearly synonymous with the term *exogeosyncline* are such descriptive expressions and passages as *molasse trough* and *deposits of a foredeep* (*or marginal deep*) *on the mobile shelf and adjoining craton.*

New England Salient

DEFINITION AND GEOGRAPHIC RELATIONS

The Appalachian orthogeosyncline swings through a wide bend from a north to northeast trend in northern New England and adjacent parts of Quebec (Pl. 3; Kay, 1951, Pl. 1). The axis of this bend projects southeastward across New England from near northern Lake Champlain; hence the bend is referred to here as the New England salient. This term applies not only to the lower and middle Paleozoic orthogeosyncline, but to secondary geosynclines and other structures in the orthogeosynclinal belt (Pls. 1, 3; Keith, 1923b, p. 313–314, Pl. 4; 1932, p. 363–364; 1933, p. 52). The areal pattern of the New England salient is also reflected to the northwest on the craton. There, lower Paleozoic (lower Middle Ordovician) rocks were deposited in an embayment whose axis coincides with the Ottawa Valley between the uplands of the Laurentian Mountains in Quebec and the Adirondack Mountains in New York (Hofmann, 1963, Fig. 10, p. 293).

Domes cored by bodies of anorthosite and related rocks in the Precambrian basement near the southeast edge of the craton form a pattern that roughly follows the bends of the orthogeosyncline not only through the northwesterly convex bend of the New England salient, but through a bend convex to the southeast in the Gaspé-Newfoundland-Labrador region. (*See* Kranck, 1961, Fig. 1; Osborne and Morin, 1962, p. 125.) Large, predominantly Paleozoic plutons occur widely in the orthogeosynclinal belt on the southeast side of the New England salient (Keith, 1923b, p. 321–322).

TRANSVERSE GEOSYNCLINAL TROUGH

A transverse geosynclinal trough near the axis of the New England salient southeast of the Vermont-Quebec geanticline (Pl. 3, between sections *D-D′* and *E-E′*) contains as much as 80,000 feet of strata in the combined eugeosynclinal zone and southeastward overlapping miogeosynclinal zone. (*See also* White and Jahns, 1950, p. 190–191.) Here the lower Paleozoic eugeosynclinal section is at least five times as thick as the equivalent miogeosynclinal section. Thirty miles northeast and southwest of the axis of the salient, this very thick section converges to less than 50,000 feet (Cady and others, 1963, p. 11, 14, 15, 18, 22, 26, 27, 32, 33; Doll, 1951, p. 21, 23, 26, 34; Ern, 1963, p. 24–25, 28, 31–32, 33, 35–36, 38,

40, 42, 44, 45, 47; Osberg, 1952, p. 22). At yet greater distances from the axis, at least to the south, the section thins to much smaller amounts (Thompson, 1952, p. 38–41); to the northeast in Quebec the thinning is countered by addition to the top of the lower Paleozoic (Pierre St. Julien, written commun., 1965). Thinning appears to be by convergence toward the flanks of the salient, inasmuch as the several major formational units involved, all of which persist across the salient, are grossly time-stratigraphic in the cross-salient direction (parallel to the orthogeosynclinal trend).

The New England salient is thus demonstrably the site of subsidence that is greater than in adjoining areas. Its apparent continuity on the craton suggests that it is basically part of a major downwarp that first appeared in Precambrian time and trends northwest across the southeastern margin of the craton.

FEATURES OF THE BEDDED ROCKS IN THE NEW ENGLAND SALIENT

Minor stratigraphic units are most varied, most abundant, and most voluminous where the salient crosses the eugeosynclinal zone in north-central Vermont. Thick sections of mafic volcanic and carbonaceous pelitic parentage are most common. They grade laterally along strike, at least to the northeast and south, into nonvolcanic and less carbonaceous schists. Nearly all the volcanic rocks in the New England salient in north-central Vermont are mafic, whereas near the northeast flank of the salient in Quebec, felsic as well as mafic volcanic rocks are common although less abundant (Duquette, 1960a, p. 4–5; 1960b, p. 4; 1961, p. 4–5; St. Julien, 1961b, p. 7; 1963b, p. 7; St. Julien and Lamarche, 1965, p. 4). Felsic volcanic rocks are also fairly common to the southwest of the salient in central and southern Vermont (Currier and Jahns, 1941, p. 1494; Skehan, 1961, p. 146). To the southeast in New Hampshire, even at the axis of the salient, felsic volcanic rocks are about as common as the mafic (Billings, 1956, p. 18, 20, 28, 30).

Early Folds and Slides

INTRODUCTION

The early folds in the orthogeosyncline are chiefly similar folds (Van Hise, 1896, p. 600),[15] in which the beds are thinner on the limbs than at the axes. They involve only the lower and middle Paleozoic bedded rocks and not the Precambrian basement. These folds were formed in both the lower and the middle Paleozoic. The stratification of the bedded rocks is the form surface in the early folds. Slides, major recumbent folds, intrastratal diapiric intrusions of bedded rocks, and less commonly upward-facing folds[16] with nearly vertical axial surfaces, are among the early structural features. Notwithstanding their large size, the slides and recumbent folds are not readily recognized. This is because they are obviously shown by the pattern of regional geologic maps (scale 1:250,000) only in limited areas, and sedimentary structures indicating tops of beds are commonly masked by metamorphism. Moreover, the slides and recumbent folds were redeformed in episodes of doming and arching, and of regional folding in a late-fold regime (to be discussed). The limbs of the early folds commonly approach parallelism with the axial surfaces. This makes the cleavage widely subparallel to bedding and accounts for a fairly widespread virtual bedding schistosity that coincides locally with micaceous minerals mimetically crystallized along bedding. (*See also* de Sitter, 1964, p. 273–277.) Most of the early folds face west and northwest toward the cratonal margin of the orthogeosynclinal belt. (*Compare with* Freedman and others, 1964, p. 627–629).

The axial surfaces of the early folds are raised in younger domes and arches and flexed into late longitudinal folds that follow the northeast trend of the orthogeosyncline. The domes and arches and longitudinal folds are the features that are principally responsible for the pattern of regional geologic maps. Most clearly identifiable as early folds are minor folds upon which minor late folds are superimposed, all within a single outcrop.

The early folds form three differently oriented sets, each of which includes major and minor folds. (1) *Cross folds* and related rodding trend northwest at about right angles to the longitudinal structural trends of the region and subparallel to the axis of the New England salient. (2) *Longitudinal folds* are rudely parallel to the northeast structural trends. Related to the longitudinal folds and similarly oriented are slides, intrastratal diapiric folds, and also syntectonic bodies of ultramafic rock intruded in the solid state. (3) *Oblique folds,* at much less than right angle to the northeast structural trends, nearly coincide with the flanks of the New England salient and swing into continuity with less distinguishable longitudinal folds that are both northeast and south of the salient and at its axis.

CROSS FOLDS

The cross folds, chiefly minor folds traceable within a single outcrop, are mostly in the lower bedded rocks (Cambrian) of the lower Paleozoic eugeosynclinal zone southeast of the Vermont-Quebec geanticline in the New England salient (Albee, 1957; Cady and others, 1962; Cady and others, 1963, p. 60–61; Chidester, 1953; 1962, p. 24–25; Christman, 1959, p. 60–62; Christman and Secor, 1961, p. 58; Clark and Eakins, 1968, p. 172; de Römer, 1961, p. 272–273; Green, 1968, p. 1631; Lamarche, 1965, p. 194, 195–197; Osberg, 1965, p. 243; Rickard, 1965, p. 526, Fig. 2B; St. Julien, 1963a, p. 256, 259–260; 1967, p. 42–46). They are also reported, though less commonly, south of the New England salient on the east slope of the Green Mountains in southern Vermont (Brace, 1953, p. 75, 94, 106) and in western Massachusetts (Hatch and others, 1967, p. 11). Moreover, the author has observed a few cross folds northeast of the salient in the Notre Dame Mountains of Quebec. Several folds large enough to show on a regional map and probably related, at least in part, to the minor cross folds, trend across the regional structural trends parallel to and a little north of the axis of the salient in northern Vermont (Cady and others, 1963, p. 56–57, 64). Deformed in these major cross folds are metasedimentary and metavolcanic strata and concordant bodies of ultramafic intrusive rocks.

The cross folds are most commonly formed of the quartzite beds in schist (Hazens Notch Formation), preserved only as isolated crests of tight folds from which the limbs have been separated. Where the fold crests are not recognizable, all that remains is a rodding parallel to the fold axes. No consistent movement sense has been detected in the cross folds; more detailed studies may or may not reveal regularity. The axial surfaces and rodding are commonly flexed by minor folds and are cut by axial-plane cleavage of the late fold regime.

LONGITUDINAL FOLDS AND SLIDES[17]

Distribution

The early longitudinal folds are most extensive in the eugeosynclinal zone southeast of the Vermont-Quebec geanticline. Early folds northwest of and on the Vermont-Quebec geanticline, notably those south of the New England salient, in and near the Taconic klippe,[18] and those that cross the axis of the salient and connect with the oblique folds on its flanks are logical members of the early longitudinal fold set. Some not yet recognized possibly exist northeast of the New England salient in Quebec. They are fairly evident in the Notre Dame Mountains at a considerable distance northeast of the salient (Béland, 1957, p. 30–31; 1962, p. 16).

Minor Longitudinal Folds

Longitudinal folds that are most clearly of the early regime are minor folds; most trend northeasterly, plunge gently in that direction, and face northwest; minor folds that face upward are found locally on the Vermont-Quebec and Somerset geanticlines. The longitudinal folds are in both lower and middle Paleozoic bedded rocks. (*See* Bain, 1931; Brace, 1953, p. 96; Christman, 1959, p. 61, Pl. 17; Clark and Eakins, 1968, p. 167–171; Dennis, 1964, p. 37; Eric and Dennis, 1958, p. 32–35; Ern, 1963, p. 58–60; Goodwin, 1963, p. 85; Green, 1968, p. 1630; Green and Guidotti, 1968, p. 264; Hall, 1959, p. 63–64; Hatch and others, 1967, p. 11; Johansson, 1963, p. 54–56; Konig, 1961, p. 64; Lamarche, 1965, p. 183–185, 194–195; Lyons, 1955, Fig. 9, p. 131; Murthy, 1957, p. 51; 1958, p. 282; Rickard, 1965, p. 524–525, Fig. 1; St. Julien, 1963a, p. 256–257, 259–261, 277, 296; 1967, p. 43–46; Stanley, 1967, p. 54–56, 60; White and Billings, 1951, p. 673–675, 677–679; White and Jahns, 1950, p. 201–207; Zen, 1961, p. 322–323; *compare also* with Freedman and others, 1964, p. 635, Fig. 6A.) In at least two areas near the south flank of the New England salient, the trend of the early longitudinal folds swings from north to northwest (Ern, 1963, p. 58; Crosby, 1963, p. 82, 117; Voight, 1965) and appears to merge with the trend of the oblique folds.

Some of the early minor folds were described as related to the similarly oriented late folds of the east limb of the Green Mountain-Sutton Mountain anticlinorium (Cady, 1956; Chidester, 1962, p. 25; Jahns, 1967, p. 140; White and Jahns, 1950, p. 201–203). However, most of the folds (which are characterized by an axial plane schistosity) are clearly older than the anticlinorial folds in which the schistosity is flexed. A few, nevertheless, are possibly in reality the late anticlinorial folds, disguised where they are more deformed and metamorphosed and their distinctive axial-plane slip cleavage grades into a schistosity. In general, the presence of minor longitudinal folds of the early regime may be inferred from cross sections in exposures in which their axial-plane cleavage (commonly a schistosity) dips at a much smaller angle than the axial-plane cleavage of the late longitudinal folds. Where the contrasting cleavage relations and the key beds traceable from outcrop to outcrop are not to be found (notably in pelitic terranes), folds such as these may be inferred only from the areal distribution and known sequence of stratigraphic units (Christman and Secor, 1961, p. 49–50).

Major Longitudinal Folds

Most of the major longitudinal folds of the early fold regime, like the minor ones, face northwest and are commonly recumbent. They are many thousand feet above the Precambrian basement.

Several major folds reported south of the south end of the Somerset geanticline are more or less clearly of the early longitudinal fold set although they are nearly as young as the late folds and are commonly included among the Acadian orogenic features. Their interpretation is controversial only because some are complexly involved with the late folds connected with the domes and arches and domal anticlines in eastern Vermont and western New Hampshire. Two are reported in middle Paleozoic rocks on the east flank of the Strafford-Willoughby arch (Hall, 1959, p. 78–79). The most obvious are in middle Paleozoic rocks adjoining the Strafford and Pomfret Domes in east-central Vermont. In this area the orientation of the axial-plane schistosity of the isoclinal early folds is closely paralleled by the axial surfaces of late folds that on the domes are also both isoclinal and characterized by an axial-plane schistosity (Chang and others, 1965, p. 54–59; Lyons, 1955, p. 123–126; Murthy, 1957, p. 51–54; 1958, p. 283; 1959, p. 582; White, 1959, p. 579–580; White and Billings, 1951, p. 683–686; White and Jahns, 1950, p. 209–218; *see also* Stanley, 1967). Another fold, the west-facing Skitchewaug nappe fronting a little west of the Connecticut River (Thompson, 1954; Trask and Thompson, 1967), is one of several (Thompson and others, 1968) preserved in synclinal tracts (of the late fold regime) in lower and middle Paleozoic rocks west of and between the Mascoma, Croydon, Unity, Alstead, Surry, Westmoreland, and Vernon domal anticlines in southwestern New Hampshire and eastern Vermont (Billings, 1955; Doll and others, 1961). The root of this nappe appears to be in the eugeosynclinal zone 15 miles southeast of the Connecticut River.[19] West of the Skitchewaug nappe are westward-facing intrastratal diapiric folds shown by opposed rotation of garnets at the top and bottom of a relatively mobile middle Paleozoic unit (Rosenfeld, 1960, 1965, 1968, p. 194–195, 196–201).

The northwestward-facing recumbent folds in northwestern New England contrast with a southeastward-facing fold of the same sort in eastern Connecticut (Dixon and Lundgren, 1968, p. 221–225; Dixon and others, 1963; Robinson, 1967a, p. 38, Fig. 4, p. 41). The great extent of the latter suggests that others may eventually be recognized throughout the southeastern New England area, especially in east-central Massachusetts and south-central New Hampshire. The roots of the southeast-facing folds appear to be a little southeast of the root of the Skitchewaug nappe discussed above. (*See* Appendix A.)

The presence of a major early longitudinal fold, or slide in the belt of middle Paleozoic rocks between the Somerset and Vermont-Quebec geanticlines, extending at least 65 miles north-northeast across the New England salient, has been proposed (Chang and others, 1965, p. 59–62; Eric and Dennis, 1958, p. 61; Ern, 1963, p. 67–68, 72–75; Goodwin, 1962; 1963, p. 98–103, 105–107; Woodland, 1965, p. 115–117). As conceived, this is a recumbent anticline rooted to the east, the beds of whose inverted limb make a synform in eastern Vermont.[20] The principal evidence for the inversion is graded bedding (Ern, 1963, Pl. 22, fig. 2). Recent studies have raised doubt as to the existence of the recumbent anticline (Woodland, oral commun., 1966). Map patterns that locally suggest folds, but which have been interpreted as sedimentary facies irregularities in homoclines (Cady, 1956), may indeed turn out to be early longitudinal folds.

Major longitudinal folds on the Stoke Mountain geanticline are probably cut by a Middle Ordovician unconformity (St. Julien and Lamarche, 1965, p. 12–13; *see also* Lamarche, 1967, p. 11). They are so far overturned to the northwest that their axial surfaces dip less than 45 degrees. A major longitudinal fold northwest of the Stoke Mountain geanticline (St. Julien, 1961b, p. 11–12) is truncated by the unconformity (St. Julien, 1963a, p. 30, 125, 256; 1965). The axial surfaces and the unconformity in this belt are nearly vertical. Other comparable early folds northwest of the geanticline are apparently truncated by a pre-Silurian unconformity (Ambrose, 1957, p. 164–169; Rickard, 1965, Figs. 1, 2c). Were the unconformities to be restored to their original subhorizontal positions before later folding, the early longitudinal folds would become northwest-facing recumbent folds.

On the Vermont-Quebec geanticline near the international boundary the major longitudinal folds are nearly upright (Clark and Eakins, 1968, Fig. 12–2; Dennis, 1964, p. 37, Fig. 11; Rickard, 1965, p. 524, Figs. 1, 2a). Moreover, they dominate the structure, and the late longitudinal folds are only minor features.

The Taconic slide is beneath the Cambrian and Lower and Middle Ordovician eugeosynclinal rocks of the Taconic klippe, and above autochthonous miogeosynclinal rocks of the same age west of the Vermont-Quebec geanticline (Cady, 1968a, p. 566–567) south of the New England salient. Its roots have been widely eroded from the source area of the allochthon in the geanticline, a site of tectonic denudation (Cady, 1968a, p. 565). The klippe contains west-facing folds (Zen, 1961, p. 314–321; 1967, p. 17–22; *see also* Doll and others, 1961, 1963; Zen, 1963a), the principal early longitudinal folds west of the Vermont-Quebec geanticline. Recumbent folds are also reported in the autochthon, on and south of the south flank of the New England salient (Crosby, 1964; Zen and Hartshorn, 1966, p. 3–4). Submarine slides and slide breccias are found in the autochthon near the axis of the salient, west of the Vermont-Quebec geanticline (Clark and Eakins, 1968, p. 167; Fisher, 1968, p. 30–32; Hawley, 1957, p. 68; Zen, 1968, p. 138).

OBLIQUE FOLDS

The oblique folds are in lower Paleozoic eugeosynclinal and miogeosynclinal bedded rocks. They trend obliquely across the Vermont-Quebec geanticline subparallel to the flanks of the New England salient (Pl. 1). Thus, those on the southwest flank of the salient in central Vermont, including an apparently related thrust fault or slide a little east of the geanticline (near the head of the White River), trend northwest (Crosby, 1963, p. 22, 67–70, 86; 1964; Osberg, 1952, Pl. 2, p. 83; written commun., 1960, *cited in* Doll and others, 1961) and those on the northeast flank in Quebec (near the middle course of the St. Francis River) trend northeast (Osberg, 1965, p. 242–243).

The oblique folds, commonly the principal folds of the area in which they occur, are generally larger than single outcrops. They are very rarely associated with the minor cross folds in the New England salient. However, at a place on the north flank of the salient where both types of folds occur together, the oblique folds appear younger than the cross folds (Osberg, 1965, p. 243). The oblique folds are only locally deformed by minor longitudinal folds of the late fold regime. Near the axis of the New England salient the oblique folds swing northeasterly from the flanks of the salient parallel to the Vermont-Quebec geanticline and become essentially early longitudinal folds. Northeast and southwest of the salient they also swing parallel to the geanticline and become longitudinal folds.

The oblique folds are asymmetrical and face southwest in Vermont on the south flank (Osberg, 1952, Pl. 23) of the New England salient; they are thus overturned away from the salient. The apparent southwest displacement on the previously mentioned thrust or slide on the south flank of the salient east of the Vermont-Quebec geanticline is in harmony with this fold movement. Northeast of the axis of the New England salient the oblique folds face southeast overturned into the salient (Osberg, 1965, Pl. 1, sections *A-A′* to *H-H′*)—a progressive change from the upward-facing symmetrical configuration and nearly vertical axial surfaces of the early longitudinal folds that are near the salient axis.

The largest oblique folds, like the large longitudinal folds, are in rocks several thousand feet stratigraphically above the Precambrian basement. They are especially found in the upper units (Pinney Hollow and Ottauquechee Formations and Bonsecours and Sweetsburg Formations) of the thick Cambrian sections on the southeast flank of the Vermont-Quebec geanticline in Vermont and Quebec (P. H. Osberg, written commun., 1960, *cited in* Doll and others, 1961; Osberg, 1965, Pl. 1), and in the Middle Ordovician (Middlebury Limestone) to the west in Vermont (Crosby, 1963, p. 22), but they die out lower in the section, especially close to the basement.[21] The zone of fullest development of the oblique folds on the south side of New England salient in Vermont seems to rise westward across the geanticline, from which it is mostly eroded. In Quebec, where erosion is less deep, the oblique folds are exposed at the axis of the geanticline and at about the same stratigraphic level on both sides.

INTERPRETATION OF THE EARLY FOLDS AND SLIDES

The early folds and slides were produced mainly by subhorizontal movements on gentle slopes; probably involved were a full range of materials from lightly loaded unconsolidated sediments, through diagenetic rocks to the higher grades of heavily loaded metamorphic rock. These were gravity movements of weak materials—movements predominantly laminar flow and slip with minimal flexing and thrusting, but with detachment as slides—restricted to the comparatively water-rich Paleozoic rocks above the Precambrian basement. The principal movement was northwestward toward the craton; it probably was triggered by successive episodes of compaction-diagenesis-metamorphism of the eugeosynclinal deposits, and uplift of the eugeosynclinal zone (including both geanticlines and geosynclinal troughs) at the end of each episode. The New England salient, like a southeast-opening bay, provided a basement framework that deflected, blocked, or reversed the northwestward movements to form especially the oblique folds and possibly the cross folds. The principal exceptions to this movement pattern are those in regions southeast of that of the present study; in Connecticut (Appendix A), possibly toward a craton on the opposite side of the orthogeosyncline, and in western Maine (R. H. Moench, oral commun., 1966), toward a geosynclinal trough that continued to subside.

The cross folds are difficult to explain because their trend is normal to the regional trends of the orthogeosyncline, the geanticlines, and longitudinal folds. Cross folds in the Green Mountains have been attributed to depressions trending east to west between resistant basement masses; the lowest Paleozoic rocks moved into and converged in these depressions with minor axial shortening of the regional structures (Brace, 1953, p. 94, 253; Cady and others, 1963, p. 63–67). A simpler possibility is that the cross folds formed on the sides of tonguelike folds that point northwestward in the same direction as the longitudinal folds face and whose movement, like the longitudinal folds, was to the northwest (Fig. 2). (*See* Carey, 1962, p. 128.) The lack of consistent movement sense in the cross folds seems to support this solution. So interpreted, the cross folds could be local aspects of the longitudinal and oblique folds.

The early longitudinal folds southeast of the Vermont-Quebec geanticline have been considered as normal updip (generally northwest-facing) drag folds on the southeast limb of the late, Green Mountain-Sutton Mountain anticlinorium. Early folds on the southeast limb of the Bronson Hill-Boundary Mountain anticlinorium might be similarly interpreted. The early longitudinal folds, however, are isoclinal, their axial surfaces are nearly parallel to schistosity, and most of them face northwest, not only indicating that they are older than the anticlinoria, but also suggesting that they formed in response to general northwestward movement that overran the geanticlines (Appendix A). Upward-facing early longitudinal folds on the Vermont-Quebec geanticline near the international boundary probably reflect lateral compression and blockage of the general northwestward movement; and they are the principal exceptions to the northwest-facing folds. The few early longitudinal folds reported to face southeast (Dennis, 1956, p. 63; Ern, 1963, p. 60–61; Murthy, 1957, p. 53; 1958, p. 283) are possibly minor folds on the overturned northwest limbs of the northwest-facing

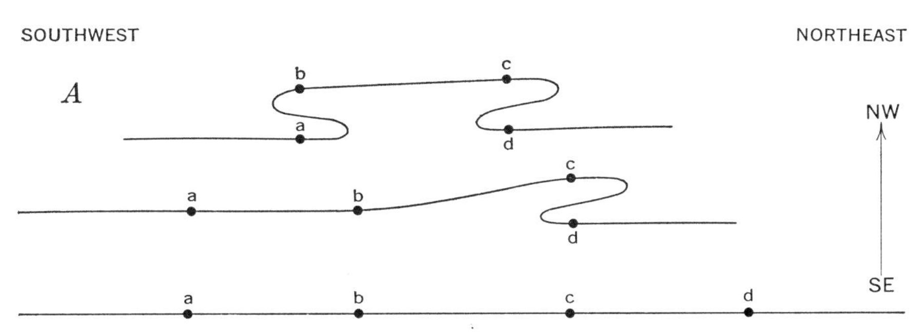

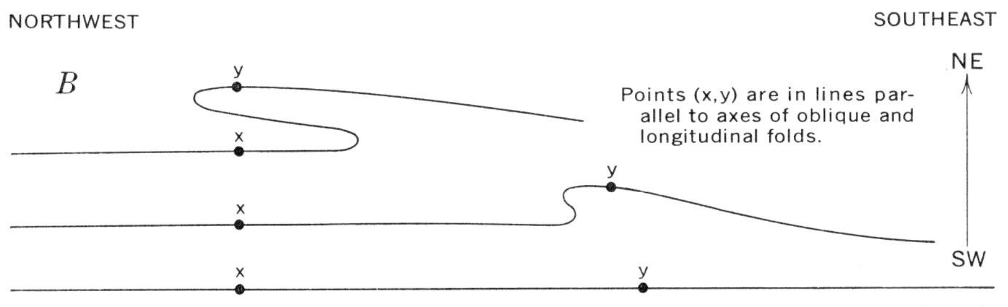

Figure 2. Diagram of early cross folds showing possible origin by convergence of northwest-trending flow lines (*modified from* Carey, 1962). A. Sections normal to axes of cross folds. B. Sections subparallel to axes of cross folds, and transecting axes of longitudinal and oblique folds.

major recumbent anticlines. These southeast-facing folds are commonly dextral as opposed to the sinistral sense of the northwest-facing folds, inasmuch as most of the folds plunge northeastward regardless of the direction they face.

The largest major folds are entirely within the upper units (Waits River and Gile Mountain Formations and Partridge, Clough, Fitch, and Littleton Formations) of the thick Ordovician to Devonian sections southeast of the Vermont-Quebec and Stoke Mountain geanticlines. Middle Ordovician pelitic rocks, some of which core the Skitchewaug nappe (Partridge Formation) and also underlie the Taconic slide (Hortonville Slate) are apparently among the most favorable movement horizons. Others involve much lower strata (Pinnacle and Underhill Formations), in the Cambrian(?) near the Vermont-Quebec geanticline and also near the axis of the New England salient. In these areas they are nevertheless well above the basement because of the greater overall thickness of the stratigraphic section. Most of the early longitudinal folds in the lower units of the Paleozoic are minor; none are known in the Precambrian basement complex. Thus early folding decreases with depth, perhaps because of basement rigidity.

The early longitudinal folds and slides formed in several different episodes (*see also* Hatch, 1968, p. D108, D112–D113). The earliest, on the northwest side of the orthogeosyncline, were younger than the cross folds and were contemporaneous with the oblique folds. Early longitudinal folds northwest of the Stoke Mountain geanticline, being truncated by a Middle Ordovician unconformity, are older than the early longitudinal folds in Silurian and Devonian rocks southeast of the Vermont-Quebec and Stoke Mountain geanticlines. The Taconic slide, which formed concurrently with the older of the early folds (Crosby, 1964; Zen, 1961, p. 326–328), overlies autochthonous Middle Ordovician (Trenton Group) strata that contain blocks spilled from the advancing slide (Zen, 1961, p. 310–313; 1967, p. 68–69; *see also* Potter, 1968). These blocks include rocks similar to Lower Cambrian quartzite, dolomitic sandstone, and dolomite in the superjacent Taconic klippe, and also slate less readily identifiable stratigraphically but probably from the klippe and including Ordovician rock. They are set in a matrix of Middle Ordovician pelites (Hortonville Slate) whose foliation extends through the slate fragments giving the fragments as well as the matrix a slaty cleavage. This foliated cleavage, which is the axial-plane cleavage of the early folds found both above and below the Taconic slide, is 460 ± 8 m.y. old according to whole-rock (K-Ar) radiometric dating (Harper, 1968a, p. 52) that independently suggests Middle Ordovician emplacement of the Taconic allochthon.

Gravity sliding and concomitant recumbent folding which would account for westward movement of the allochthonous Taconic rocks from the southern part of the Vermont-Quebec geanticline, has been variously suggested and interpreted by several authors (Cady, 1945, p. 578–579; 1967, p. 64; 1968a, p. 571–574; Crosby, 1963; p. 83–86; 1964; Platt, 1962, p. 109–110; Zen, 1961, p. 326–329, 335; 1967, p. 62–69; 1968, p. 135–137). Moreover, northwestward-facing early longitudinal recumbent folds, known at least as far southeast as the roots of the Skitchewaug nappe, well southeast of the Connecticut River (Thompson, 1954, map, p. 36; Thompson and others, 1968, Pl. 15–1a, 15–1b), indicate northwestward gravity flow and sliding more widespread and persistent in time than simply the westward movement from the Vermont-Quebec geanticline. Inasmuch as the folds face westward near the southern part of the Vermont-Quebec geanticline and at the crest of the Stoke Mountain geanticline, apparently the general northwestward movement partly surmounted the geanticlines. This overwhelming movement perhaps may stem from principal uplift in a more mobile central belt of the eugeosynclinal zone southeast of the geanticlines (Appendix A; *see also* Cady, 1968a).

The cross section of the eugeosynclinal zone, miogeosynclinal zone, and included Vermont-Quebec geanticline restored to the Middle Ordovician (Pl. 3), suggests the mechanism of the Taconic slide (*see also* Cady, 1968a, Fig. 2, p. 568). It is proposed that abundant pore and combined water was expelled during compaction and regional metamorphism from the thick eugeosynclinal rocks on the east side of the geanticline, and that this water migrated westward into the thinner eugeosynclinal section on the geanticline. Sufficient amounts arrived, despite expectable leakage on the way, to compensate for leakage from the geanticlinal section during the long Early Cambrian to Middle Ordovician episode of sedimentation. Relief on the geanticline was least compensated by sedimentation on the west flank. When this relief became sufficient during the Middle Ordovician, enough fluid pore pressure had been maintained down to the level of the Lower Cambrian to float the allochthon westward. It is especially significant in this connection that the least pervious (pelitic) rocks increase in thickness and lateral extent toward the top of the section on and east of the geanticline, thus they probably trapped water under lithostatic pressure in the more permeable rocks at depth, producing anomalously high pore pressures in the vicinity of the roots of the Taconic slide (*see also* Heard and Rubey, 1966, p. 748–749, 754–756; Platt, 1962; Rubey and Hubbert, 1959, p. 173–185; 1965, p. 471–472). Moreover, the Middle Ordovician rocks at the top of the stratigraphic sections both beneath and resting upon the Taconic slide are part of a continuous and widespread pelitic facies that probably eased movement of the overriding allochthon (Cady, 1945, p. 560, 579) especially where it included uncompacted sediments in an environment of sea water (*see* Zen, 1961, p. 328; 1967, p. 69).

If, as just suggested, anomalously high fluid pore pressures in the vicinity of the roots of the Taconic slide made it possible for the allochthon to float westward, then the radiometrically dated (460 ± 8 m.y.) Middle Ordovician foliation in and near the Taconic klippe (discussed in a preceding paragraph), which intersects biostratigraphically dated Lower Cambrian to Middle Ordovician sedimentary rocks, was probably formed in rocks compacted under load, that had their water expelled, and that were statically metamorphosed very soon after the allochthon was emplaced.

The oblique folds, as already described, are variations of the early longitudinal folds; hence much of the foregoing interpretation of the longitudinal folds applies to the oblique folds. The overturning of those on the southwest flank of the New England salient toward the southwest away from the salient, in opposition to the overturning of those on the northeast flank to the southeast into the salient, may be interpreted as follows.

The movement to the southwest, out of the salient, is consistent with the movements connected with the longitudinal folds. It took place because the eugeosynclinal zone east of the Vermont-Quebec geanticline in central Vermont, despite geosynclinal subsidence in the salient and environs, was probably both uplifted episodically and kept full of accumulations that spread westward over the geanticline. The southeastward movement into the northeast side of the New England salient was perhaps brought about because the eugeosynclinal zone southeast of the Vermont-Quebec geanticline in Quebec was not filled full enough or uplifted sufficiently to make possible northwestward movement out of the salient. The problems of different directions of horizontal movement, such as the above, are resolved by the conceiving of the New England salient as a highly mobile zone of relative extremes of subsidence and uplift. The variously directed horizontal movements in this zone may be thought of as responses to gravity potential by rocks that have undergone differential vertical movements —rather than necessarily as a response to compression directed across the orthogeosyncline.

FOOTNOTES

[15] The similar folds of Van Hise include approximately Donath's (1963) passive folds and flexural flow folds, which form under conditions of high mean ductility. (*See also* Carey, 1962, p. 98–101; Rast, 1964, p. 178–179.)

[16] "A fold faces in a direction normal to its axis, along the axial plane, and toward the younger beds. This coincides with the direction towards which the beds face [top] at the hinge. A normal upright fold faces upward. An anticline closing downwards faces downwards" (Shackleton, 1958, p. 363).

[17] "A slide is a fault formed in close connection with folding, which is broadly conformable with a major geometric feature (either fold limb or axial surface) of the structure, and which is accompanied by thinning and/or excision of members of the rock-succession affected by the folding" (Fleuty, 1964, p. 454). Moreover, Fleuty stated (p. 454–455) that slides can form along the axial surfaces as well as along limbs of folds and can have original steep dips as well as gentle dips. He further stated that slides do not include faults that break across and displace bands without appreciable attenuation and that they may or may not be gravitational gliding planes.

[18] The expressions "Taconic klippe and Taconic slide" here refer to several klippen and the slides beneath them, whose collective distribution is nearly continuous in the vicinity of the Taconic Range in western New England and eastern New York (*see also* Cady, 1968, p. 571; Zen, 1967, p. 12–13).

[19] The Skitchewaug nappe probably extends southward into Massachusetts (Robinson, 1967b, p. 23–24) where "reconstruction of early recumbent fold structures appears to account for 30 km of overburden, all presumably metamorphic rock, and suggests a 45 km depth may be reasonable," although "new hydrothermal data reported in 1966 [Newton, 1966; *see also* Althaus, 1967; Holm and Kleppa, 1966, p. 1618, 1621] indicate a pressure of 6.3 kb corresponding to a depth of 24 km under lithostatic pressure" is all that is necessary to reach the kyanite-sillimanite inversion recognized in the rocks (Robinson, 1966). Even lesser pressures and corresponding depths have been suggested (Thompson and others, 1968, p. 216).

[20] This synform coincides with the Townshend-Brownington syncline, a complex fold between the late longitudinal folds on the west limb of the Connecticut Valley-Gaspé synclinorium and the domes and arches located near the axis of the synclinorium (Doll and others, 1961). It will be discussed under the general heading of "late folds and thrust faults."

[21] The downward disappearance of the oblique folds probably explains the difficulty in differentiating them from the east limb of the Middlebury synclinorium (a late longitudinal fold) in west-central Vermont. (*See* Voight, 1965.)

Structural Relations of Ultramafic, Mafic and Intermediate Intrusive Rocks of the Eugeosynclinal Zone

Semiconcordant relations of the ultramafic, mafic, and intermediate intrusive rocks of the eugeosynclinal zone with the early longitudinal folds are fairly clear in southern Quebec, northwest of the Stoke Mountains geanticline. In that area, mafic and intermediate plutons, chiefly of gabbro and diorite (de Römer, 1958, p. 8–9; 1963; St. Julien, 1961a, p. 8; 1965) below the Middle Ordovician unconformity, are syntectonic with the folds, as are bodies of ultramafic rock (St. Julien, 1961a, Map 1363) in the gabbro, diorite, and bedded rocks beneath the unconformity .The ultramafites occur far enough west in Vermont to be involved not only with the early longitudinal folds but with the early cross folds southeast of the Vermont-Quebec geanticline. The contacts between most of the ultramafites and the bedded rocks are subparallel to the foliation in the bedded rocks and to the bedding and axial-plane cleavage of early folds. (*See* Cady and others, 1963, p. 35; Chidester, 1962, p. 17–23; 1968, p. 348; Jahns, 1967, p. 143; St. Julien, 1961a, Map 1363; 1965.) Metamorphosed mafic dikes are likewise subparallel to the foliation, and to the bedding and axial-plane cleavage of the early folds, though their contacts locally transect (Jahns, 1967, Fig. 5.3), and in many places are transected by the axial plane cleavage and foliation (Billings and White 1950, p. 632; Moore, 1949, p. 1639; White and Billings, 1951, p. 662–663; White and Jahns, 1950, p. 194). These dikes, as well as the ultramafic bodies which cut them, are flexed by folds of the late regime, and will be discussed under a later heading (Jahns 1967, p. 143).

The deformational style of the ultramafites contrasts markedly with that of the minor early folds (similar folds) in the bedded rocks. Especially characteristic of the ultramafic rocks are shear polyhedrons—irregular polyhedral units of massive, essentially unsheared serpentinite separated by thin and locally conjugate zones of sheared serpentinite (Chidester, 1962, p. 24, 71; 1968, p. 348; Chidester and others, 1951, p. 7–8; Jahns, 1967, p. 146). Mesh structure characterizes the serpentinite in the shear polyhedrons, whereas the intervening sheared serpentinite shows a foliate structure. Intersecting joints are characteristic where dunite and peridotite are common (Cady and others, 1963, p. 62).[22] Inclusions of wall rocks contain tight folds that were formed before (Jahns, 1967, Fig. 5.3; White and Jahns, 1950, p. 196), or possibly penecontemporaneously with emplacement of the ultramafic rocks. The strangely brittle style of the ultramafic rocks is probably to be explained by their initial isotropy, as compared with the anisotropy of the containing bedded rocks. (*See also* Lapham and McKague, 1964, p. 656.)

Outer zones of some of the ultramafic plutons near the contacts are made up of schistose serpentinite, whose fabric is changed entirely from mesh to foliate. In these zones, the schistosity parallels both the contacts and the schistosity of the adjoining regionally metamorphosed bedded rocks and obliterates the shear polyhedrons and related structures (Cady and others, 1963, p. B40-B41; Jahns, 1967, p. 145–146). The deformational style of the outer zones thus agrees with that of the bedded rocks and probably reflects an increase in ductility during serpentinization to values equal to or greater than the ductility of the bedded rocks.[23]

The ultramafic rocks evidently arrived at their present sites in the solid and relatively cool state that was their condition during most of their transport (Cady and others, 1963, p. B45-B46; Chidester, 1962, p. 89; 1968, p. 350, 353; Chidester and others, 1951, p. 7–8; Jahns, 1967, p. 156). They evidently were transported in a northwestward and upward movement in the eugeosynclinal section, converging upward on the southeast flank of the Vermont-Quebec geanticline. Comparatively plastic serpentinized dunite and peridotite moved chiefly parallel to bedding and foliation in response both to pressures of the thickly accumulated overlying strata in the geosynclinal trough deposits and possibly to tectonic overpressure directed from the southeast. The mobility of the ultramafic rocks apparently increased as serpentinization progressed, ductility of the rocks increased and density decreased. On the other hand, plutons of mafic and intermediate rocks (mainly gabbro and diorite) which are cut by the cold intruded ultramafic rocks and are less widely distributed, lost their mobility when they crystallized from magma, so were less active participants in the early longitudinal folding (*see also* de Römer, 1963; Riordon, 1954, p. 12; St. Julien, 1963a, p. 198, 206, 216).[24]

The restricted distribution of the gabbro and diorite, and to a somewhat less extent the dunite and peridotite, is illustrated by the wide failure of these rocks to crop out in the thick section of the New England salient. In that section, only the tectonically more mobile serpentinized ultramafites were transported far enough northwest to reach rocks exposed in the Green Mountains and vicinity. It seems unlikely that the westernmost ultramafic rocks on the northwest side of the Green Mountains are rooted to the west or transported from that direction, because to the west these rocks decrease in abundance both laterally and at stratigraphic depth as the Vermont-Quebec geanticline and the miogeosynclinal zone are approached.

Ultramafites began to be injected at low regional metamorphic grades before the foliation was formed. Injection stopped when and where metamorphism was sufficiently complete that water was not readily available for serpentinization.

After that and during continuing regional metamorphism the serpentinite was replaced by successive talc-carbonate and carbonate-quartz rocks. The bedded rocks at the contacts were first altered—at low levels—to serpentine-chlorite rock (including free carbon) and to rodingite in the presence of the dry and perhaps hot dunite and peridotite. At the higher levels to which the serpentinized ultramafites ascended, bedded rocks near the contacts under high partial pressure of carbon dioxide, exchanged materials with the serpentinite to form steatite, blackwall chlorite rock, and albite porphyroblast rocks (Appendix D).

The wide failure of ultramafic rocks to transect upper stratigraphic levels (Silurian and Devonian, *see* footnote 11, p. 31) is possibly due to metamorphism at lower levels (Ordovician), even before deposition of the upper bedded rocks. That is, the rise of later generations of ultramafic rocks to the upper levels was possibly prevented by the water-poor condition of the lower rocks through which they would have had to pass.

Intrusion of the nearly concordant mafic dike rocks is believed to have taken place principally a little before but also to some extent a little after foliation formed.

FOOTNOTES

[22] The intersecting joints and shears in the ultramafic rocks in some places transect layering that originated before serpentinization in the dunite and peridotite (Cady and others, 1963, p. B36, B48, B59; Chidester, 1962, p. 17). The layering is possibly a relic of crystal accumulation from magma, or of passive flow of the otherwise isotropic crystalline rock, or both. If passive flow of the crystalline rock is involved, it would seem that possibly at sufficient depth under high confining pressures and high temperatures, the dunite and peridotite are analogous to diapiric salt and more ductile than the bedded rocks (*see also* Raleigh, 1967, p. 197).

[23] It appears that were the adjoining bedded rocks to provide the water necessary for serpentinization, and were the necessary deforming stresses to continue, the deformational style would eventually be in complete agreement with that of the country rock. That these styles rarely are in complete agreement is probably to be attributed chiefly to early drying up of the bedded rocks during metamorphism. This is an event that probably not only cut off the source of water for serpentinization, but that also reduced the ductility of (stiffened) the enclosing bedded rocks, thereby bridging deformational stress around the ultramafic rocks.

The tectonic mobility of the serpentinite may have been locally enhanced by increased temperature. Experimental thermal dehydration of serpentinite (Raleigh and Paterson, 1965), which resulted in the formation of forsterite and talc, "embrittled and weakened serpentinite." This is "attributed to a reduction in the effective confining pressure due to the pore pressure of water released during dehydration and to a loss in cohesive strength due to changes in structure upon dehydration." Such dehydration does not appear to have been extensive here, however, inasmuch as olivine has not been found as a replacement for serpentinite and there is rarely evidence in the adjoining metamorphosed bedded rocks that the critical temperature of 500° C was attained. Experimental dehydration and weakening of serpentinite under shear stress commenced at temperatures well below 500° C, but the confining pressures necessary were more than 15 kb (Riecker and Rooney, 1966; *see also* Sclar and Carrison, 1966; Rooney and Riecker, 1966, and Jahns, 1967, p. 156).

[24] Small mafic dikes are known to cut the ultramafic rocks in northern Vermont (Cady and others, 1963, p. B49), but the larger gabbro bodies found in that area are very likely cut by cold intruded ultramafic rocks, as in adjacent Quebec.

Concordant Calc-Alkalic Plutons

Largely concordant though locally discordant, and commonly foliate lower and middle Paleozoic granitic plutons mostly of batholithic dimensions, are emplaced at several levels chiefly in the bedded rocks, early folds, and slides of the lower and middle Paleozoic eugeosynclinal zone (Pls. 1 and 3). They are synkinematic intrusives; the contacts of which are nearly parallel to bedding, to axial surfaces of the folds, and (less commonly) to foliation. They are calc-alkalic and mainly intermediate to felsic in composition (Billings and Keevil, 1946, p. 810–811; Cady and Chidester, 1957).

The lowest known plutons, in terms of enclosing strata, are to the northwest and the highest to the southeast. Thus their horizontal and vertical distribution, like that of the ultramafic to intermediate rocks deformed in the early folds, reflects the southeastward offlap of the eugeosynclinal zone. Moreover, some of the concordant plutons are known to be truncated upward by unconformities reported at sucessively higher levels in the direction of southeastward offlap. These relationships are as follows.

Granitic dike rocks (Cooke, 1937, p. 76–84; Riordon, 1954, p. 12) between the Vermont-Quebec and Stoke Mountain geanticlines are emplaced in ultramafic rock (*see also* Appendix I). Small Lower Ordovician(?) albite granite and albite rhyolite plutons (Duquette, 1960b, p. 4–5, Map 1344; Lamarche, 1967, p. 9–10; St. Julien, 1963b, p. 10, Map 1466; 1965; St. Julien and Lamarche, 1965, p. 10–11) near the Stoke Mountain geanticline in Quebec, cut Lower or Middle Cambrian and Middle Ordovician rocks and are apparently truncated by a Middle Ordovician unconformity (St. Julien, 1963a, p. 205). (*See* Pl. 3, section *C-C′*.) Foliation transects their contacts (St. Julien, 1963a, p. 200–203). Upper Middle or Upper Ordovician granodiorite, quartz monzonite, and granite plutons (Highlandcroft Plutonic Series, Pl. 3, section *D-D′*; Mascoma Group of Naylor, 1967, p. 24–26-"unstratified core rocks" of Naylor, 1968, p. 233–234; 1969, p. 411–412, not shown on Pl. 3) are exposed near the Somerset geanticline and at the cores of domal anticlines ("Oliverian" domes, Billings, 1955) to the south of the Somerset geanticline. They cut Middle Ordovician strata and are truncated by an unconformity beneath the Silurian (Albee, 1961, p. C51, Fig. 168.1; Billings, 1955, 1956, p. 46–48, 121–122, 148; Cady, 1960, p. 565; Doll and others, 1961; Naylor, 1967, p. 24–28, 67–80; 1968, p. 233–234, 238–239;

1969, p. 419–420). Their contacts are transected by foliation (Eric and Dennis, 1958, p. 42, 46). Relatively large plutons of Lower(?) Devonian granodiorite and quartz monzonite (Bethlehem Gneiss and Kinsman Quartz Monzonite—Billings, 1955, 1956, p. 55–57, 58–61, 125–127; Doll and others, 1961, section *E-E′*; Page, 1968, p. 378) and related pegmatites (Page, 1968, p. 378–380) are emplaced in predominantly pelitic middle Paleozoic eugeosynclinal bedded rocks (Littleton Formation) that are involved in early folds southeast of the geanticlines. (*See* Thompson and others, 1968, p. 208–209[25], and Pl. 3, sections *E-E′*, *F-F′*.) Their internal foliation (most notably that of the Bethlehem Gneiss) is about parallel to the contacts and to the foliation of the adjoining country rock (Billings, 1937, p. 536–538; Chapman, 1939, p. 161–164; 1952, p. 406–407, 408; Hadley, 1942, p. 164–166; Heald, 1950, p. 70–74; Page, 1937, p. 81–83, 91; 1968, p. 378).

The Lower(?) Devonian concordant calc-alkalic plutons in western and northern New Hampshire (Mount Clough, Cardigan, and Lincoln plutons—Billings, 1955) coincide widely with the sillimanite zone of regional metamorphism. Rocks of sillimanite grade are of Ordovician, Silurian, and Devonian age, as already mentioned. It should be emphasized that the sillimanite metamorphic zone is not localized by the plutons, thus the plutons formed under conditions of pressure and temperature harmonious with those of the adjoining regionally metamorphosed sillimanite schists. They were emplaced penecontemporaneously with development of the foliation (*see also* Thompson and Norton, 1968, p. 235).

FOOTNOTES

[25] The Bethlehem Gneiss, Kinsman Quartz Monzonite, and various other plutonic rocks that are emplaced mainly in rocks that have been assigned to the Silurian(?) and Devonian Littleton Formation or its equivalents in New Hampshire, comprise the principal units of the New Hampshire Plutonic Series (Billings, 1956, p. 53–65; Cady, 1960, p. 565). Their characteristic foliation and concordance contrast with features of various other units (typically the Concord and Bickford Granites) that have been included in the New Hampshire Plutonic Series, whose rocks are commonly massive, form discordant plutons, and some of which cut the concordant plutons. Because of these divergent tectonic relations the concordant and discordant granitic rocks are here considered separately. Goldsmith (1964) compiled a manuscript geologic map of New England upon which "syn- and post-tectonic" (concordant and discordant) "calc-alkalic granites" of the New Hampshire Plutonic Series are distinguished by their patterns, the post-tectonic being random. This map has proved useful in tracing the distribution of the concordant and discordant granitic rocks and their metamorphic relations in outlying regions.

Regional Foliation

USE AS REFERENCE SURFACES

The regional foliation, which is subparallel to both the axial surfaces and the limbs of the early folds, is the most nearly ubiquitous and pervasive structural element useful as reference surfaces in the analysis of structural features found in the eugeosynclinal zone. It is most commonly a schistosity, less commonly a slaty cleavage or gneissosity, and where genetic relations to the bedded rocks are clear, it is recognized either as parallel to bedding, or as an axial plane foliation (including slip-cleavage schistosity) (Fairbairn, 1935, p. 594–597, 607–608). Where the genetic relations of the regional foliation to the bedded rocks are not clear, as is fairly commonly the case, the regional structure is conveniently described, through use of the foliation instead of bedding as the form surface in structural elements such as folds, slip cleavage, and faults – without inference as to the genesis of the foliation. It should be pointed out that this procedure is complicated locally, especially in micaceous rocks. This procedure is also complicated in highly metamorphosed granular rocks where slip cleavage that deforms the regional foliation grades into a transverse foliation which is commonly a schistosity. This transverse foliation obliterates the regional foliation but is difficult to distinguish from the regional foliation. (*See also* Cady and others, 1963, p. 57–62; White, 1949.) The foliation changes more or less abruptly into slip cleavage and even into fracture cleavage, shear joints, and bedding joints where it encounters granular rocks.

RESTORATION TO ITS POSITION BEFORE DEFORMATION

The regional foliation, like the bedding, provides especially useful surfaces of reference as restored to its position before it was deformed. Restoration involves elimination of all of the later formed major structural features, such as domes and arches, anticlinoria and synclinoria, and thrust faults responsible for the geologic map pattern (Billings, 1955; Doll and others, 1961; Quebec Department of Mines, 1957). After such a restoration, most of the early folds remain. These are tight minor and major folds, many of which are recumbent, subparallel to whose limbs, axial plane cleavage, and slip cleavage the regional foliation is formed. Inasmuch as the early minor folds are sparse and most of them are dismembered and have dimensions rarely larger than an outcrop, the regional foliation generally approaches parallelism with bedding. Thus restored, the foliation conforms widely to the gentle configuration of the geosynclinal troughs and intervening geanticlines (Pl. 3). The foliation, and with it the bedding, tabular intrusive bodies, recumbent folds, cleavage, and slides, diverges laterally toward the troughs, including the transverse trough of the New England salient. This divergence indicates subsidence of the troughs, which is greatest near their axes, and in turn a possible control of the direction of transport, not only of sediments and lava, but of slides and recumbent folds. In rare places, where the axial surfaces of the early folds are vertical, the regional foliation is nearly vertical and less nearly parallel to the bedding on the limbs of the folds. This is possibly due to upward movements caused by deflection of recumbent folds against barriers, including other folds, as well as to lateral compression.

The regional foliation masks the surface of the Taconic slide; this relationship is consistent with the observation (Zen, 1960, p. 132) that the Taconic allochthon predates at least the climax of metamorphism. The Taconic slide zone is, with a few exceptions (Zen, 1961, Pl. 3, fig. 4), in slates and phyllites (Cady, 1945, p. 559, 569; Zen, 1961, p. 313) in which the regional foliation is formed. The slide, along with the foliation, is flexed in late longitudinal folds (Zen, 1964b, p. 64) and is cut by the late longitudinal thrust faults rooted in the autochthonous rocks beneath the slide (Thompson, 1959, p. 72) that together are mainly responsible for the map pattern of the Taconic klippe (Doll and others, 1961). The fact that the regional foliation masks the rootless Taconic slide suggests that it also masks the slides beneath less detached allochthons that have gone undetected because their stratigraphic and structural relations are not completely clear.

RELATION TO REGIONAL METAMORPHISM

The regional foliation is closely related to the lower grades of regional metamorphism. The minerals that accentuate the foliation are sericitic muscovite, less commonly paragonite, fine-grained chlorite and biotite, of which the latter two are particularly diagnostic of the lower grades. Porphyroblasts of albite and less commonly of other minerals transect the foliation; most contain relic foliate minerals and are therefore superimposed upon the foliation, and in many instances upon folded and sheared foliation (de Römer, 1961, p. 278–279; Osberg, 1952, p. 94, 95; Rickard, 1964a, p. 90, 1965, p. 527). In the higher grades the sericite is enlarged, maintaining the orientation of the regional foliation, but the chlorite and biotite disappear and porphyroblastic rather than foliate minerals, including garnet, kyanite, staurolite, and sillimanite take their place (Cady and others, 1962). Some of the garnet shows simultaneous growth and rotation (Rosenfeld, 1968)—rotation that began during the early, passive folding, and continued with development of the regional foliation as both axial-plane and bedding schistosity, and in some places was reversed during doming and late folding.

It seems fairly clear, as already mentioned, that the regional foliation first appeared as a feature of, and is chiefly a relic of, low-grade regional metamorphism; moreover, in the pelitic rocks it may be a relic of compaction of unconsolidated sediments. It began to form near the bottom of geosynclinal deposits and progressed upward in the deposits concomitantly with their accumulation. This seems especially clear in Quebec where two unconformities—that beneath the Middle Ordovician and that at the base of the Silurian (middle Paleozoic)—mark sharp upward decreases of the foliate condition of the bedded rocks. (*See* Clark, 1937; Riordon, 1957.) The gross effect is of several and mostly nonorogenic initial metamorphic episodes spaced at widely different times from Cambrian to Permian.

Domes and Arches

A longitudinal tract of middle Paleozoic domes and arches, some of which expose Precambrian basement rocks, extends diagonally northeastward from a little west of the axis of the Vermont-Quebec geanticline in southern Vermont. It crosses the geanticline in south-central Vermont and apparently terminates in northeastern Vermont between the Stoke Mountain and Somerset geanticlines (Pl. 1). The domes and arches contain Paleozoic bedded rocks of the interval Lower Cambrian through probable Lower Devonian.

Drag folds characteristically face downdip on the flanks of the domes and arches, as opposed to the folds that face updip on the limbs of normal anticlines. These reverse drag folds are similar folds (Van Hise, 1896, p. 600); therefore, they have much the appearance of the tight early folds already discussed. However, the late reverse drag folds, unlike the early folds in which the regional foliation is subparallel to the axial surfaces, are folds in the regional foliation, which is their proper form surface (White, 1949, p. 590; White and Jahns, 1950, p. 203–206, 208, 209–210). Where deformation connected with the domes and arches is greatest, especially near their crests, the regional foliation is obliterated by a new foliation. This foliation is parallel to the axial surfaces of the reverse drags, which are therefore difficult to distinguish from the early folds. Consequently, a controversy (already cited in the section on early folds and slides) has arisen about the relative extent and significance of the early and late folds on the domes and arches.

The Strafford-Willoughby arch at the northeast end of the tract (Doll and others, 1961; Eric and Dennis, 1958, p. 10–31) is the most extensive arch. This arch is symmetrical to the axis of the New England salient, which it crosses at approximately a right angle; it extends for about 40 miles both northeast and southwest. Because the arch is in extremely thick Paleozoic strata it fails to show a core of Precambrian basement rocks. To the southwest, on the east flank of the Vermont-Quebec geanticline, the Paleozoic bedded rocks thin and the Precambrian basement is exposed in several domes, most notably the Chester dome, that are separated by saddles. These domes extend across the geanticlinal axis into western Massachusetts (Christensen, 1963, p. 106; Skehan, 1961, p. 110–118, Thompson, 1952, p. 20, 21) and probably on into western Connecticut in the Berkshire Highlands (Emerson, 1917, p. 10, Rodgers and others, 1956).

The reverse drag folds on the flanks of the domes and arches indicate rather clearly that the domes and arches were raised chiefly by vertical upward pressures from beneath their centers (Dennis, 1956, p. 62–63, 67–68; Ern, 1963, p. 68–70; Eric and Dennis, 1958, p. 31, 53–54; Goodwin, 1963, p. 103–105, 106; Hall, 1959, p. 64, 89; Konig, 1961, p. 75; Lyons, 1955, p. 125; Murthy, 1957, p. 56–66, 69; Rosenfeld, 1968, p. 192–193; Skehan, 1961, p. 110–130; Thompson, 1952, p. 20, 21; Thompson and Rosenfeld, 1951; White and Jahns, 1950, p. 218–219). Gravity study of the Strafford and Pomfret domes near the south end of the Strafford-Willoughby arch indicates the presence of low-density rock at shallow depth beneath these structures (Bean, 1953, p. 529–532). The ascent of these domes possibly "became the dominant late stage tectonic event" (Lyons, 1955, p. 125). This low-density rock is probably contained in discordant granitic plutons, such as those that crop out widely and coincide with gravity lows farther north in the Strafford-Willoughby arch (Diment, 1953, Pl. 1; 1968, Fig. 30–3, p. 402). These plutons transect the folds and foliation of the domes and arches and it therefore seems unlikely that they are primarily responsible for the doming, although they may have locally modified the structure of the bedded rocks (Konig, 1961, p. 75; Myers, 1964, p. 57–58). Instead, a more general buoyancy is possibly accountable. (*See also* Thompson, 1952, p. 21.) Metamorphism has been proposed as a cause (Naylor and Wasserburg, 1966) although it seems quite as likely that shortening of the cross section of the orthogeosynclinal belt caused upward squeezing of basement rocks to form the domes and arches. (*See* the discussion of buoyancy versus squeezing in Appendix A, in another although possibly related context.)

Late Folds and Thrust Faults

INTRODUCTION

The late folds are grossly parallel folds in which by definition the beds are about as thick on the limbs as at the axes (Van Hise, 1896, p. 600)[26]; the axial surfaces of these folds are commonly at an angle to the limbs. They include the principal northeast trending major longitudinal folds that involve the Precambrian basement and also steeply plunging minor folds that involve rocks above the basement. All are middle Paleozoic structural features in which the most commonly observed form surface in the eugeosynclinal zone is the regional foliation, and in the miogeosynclinal zone is the bedding. The late folds commonly show normal drag-fold relationships—the smaller folds on the limbs of the larger folds face in the updip direction. The thrust faults range in trend from north to northeast and parallel the folds. The late folds are commonly not confused with the early longitudinal folds, already described, because of their distinctive flexural style; moreover, the minor folds of the late regime are the folds usually seen in outcrop.

The late folds and thrust faults, with some exceptions, dominate both maps and structure sections (Pl. 1). The late major folds and thrusts determine the patterns of the regional geologic maps (scale 1:250,000), except in geanticlinal tracts where only the larger of the major folds that include broad flexures of the contact with the Precambrian basement dominate the major folds of the early regime. The minor late folds and their related axial plane cleavage are more commonly recognizable than the minor early ones. This is because the limbs of the late folds less frequently approach parallelism with their axial surfaces, and as a corollary, the related axial plane cleavage is rarely parallel to bedding. (*Compare with* Freedman and others, 1964, p. 629–630.)

The axial-plane cleavage of the late folds ranges from fracture, through crenulation cleavage (Rickard, 1961, p. 329–330), to slip cleavage and slip-cleavage schistosity. It generally dips much more steeply than the axial-plane cleavage of the early folds. The fracture cleavage commonly transects fairly competent nonfoliate beds at a relatively high angle to their surfaces and may also be at considerable angles to the axial surfaces. The crenulation cleavage, including kink layers,[27] transects the regional foliation and is nearest to parallelism with axial surfaces in thinly interbedded incompetent and competent foliate strata. The slip cleavage and slip-cleavage schistosity also transect the regional foliation in predominantly incompetent foliate beds; in such rocks they are commonly subparallel to both the axial surfaces of the folds and to the bedding on the limbs.

Although the late folds are grossly parallel folds, their style changes with lithology, thickness of overlying strata, and amount of deformation—changes most marked in the transition from the relatively brittle environment of the miogeosynclinal zone to the more plastic environment of the eugeosynclinal zone. In the northwestern part of the miogeosynclinal zone, the folds can be classified most strictly as parallel folds—flexural slip folds marked by slip between flexed layers. Southeastward in the miogeosynclinal zone, similar folds—quasi-flexural folds and flexural flow folds, the latter especially marked by flow within and parallel to the layers—are characteristic of the less competent strata. Characteristic farther southeast in the eugeosynclinal zone, are both quasi-flexural and passive-slip and flow folds—the latter two are especially marked by movements across layers of uniform competency. The passive-slip and flow folds approach the style of the similar folds already discussed under the heading of "early folds and slides." (*See* Donath and Parker, 1964.)

The axial-plane cleavage associated with the late folds varies with the competency of the rock. Fracture cleavage, which shows the widest departures from parallelism with the axial surfaces of the folds, is associated mainly with the flexural-slip folds in the miogeosynclinal zone, but is formed locally in the most competent rocks in the eugeosyclinal zone. Crenulation cleavage is associated with flexural folds, especially in the eugeosynclinal zone, where differences in competency are not extreme and are between thin layers. Slip cleavage and slip-cleavage schistosity are associated with passive-slip and flow folds in the eugeosynclinal zone; they are commonly crenulation cleavage that shows slip across the boundaries of the layers.

The layers (form surfaces—bedding or foliation, or both) in passive-slip and flow folds are offset on the axial-plane cleavage in directions both the same and the opposite to those of the nearby or coincident flexural drag folds, including crenulations (Brace, 1953, p. 88–89; Dennis, 1960, p. 572; 1964, p. 47–49; de Römer, 1961, p. 275–277; Osberg, 1952, p. 85–87; 1965, p. 230; Woodland, 1965, p. 76–81). The author has observed a predominance of offsets opposite in sense to that of the drag folds, especially to that of the drag folds that form kink layers (*see also* Clark and Eakins, 1968, Fig. 12–5; Hawkes, 1940, p. 142; Rickard, 1961, p. 328–329, Fig. 2-D), in the eugeosynclinal zone. The kink layers, a small percentage of which form conjugate (actually intersecting) sets, are relatively competent tabular units, slips at the boundaries of which show offset opposite in sense to the kinks. Conjugate kink layers that show such offset are clear evidence of pure flattening, in which the angles between the intersecting kink layers and between the kink layers and bedding foliation were both reduced. Where a single kink layer instead of conjugate kink layers is involved, and the offsets at the borders of the kink layers are opposite in sense to the kinks, there is clear evidence that the folds have been accentuated in amplitude by passive slip and flow. This process can be visualized if it is recalled that in flexural folds the axial-plane cleavage, especially fracture cleavage, commonly diverges toward the convex sides of form surfaces of folds, but if the cleavage is rotated toward parallelism with the axial surfaces, as in the passive folds, rock must move toward the fold crests.[28]

The passive folds and related slip and flow cleavages are most strongly developed in synclinal tracts between the late longitudinal folds, or between the late longitudinal folds and the domes and arches. The steeply plunging minor folds are most common in these tracts.

Mineral lineations, and less commonly slickensides and minor folds that plunge downdip on bedding and bedding foliation surfaces nearly at right angles to fold axes, are also features of the late fold regime. (Brace, 1953, p. 101–105; Cady, 1956; Crosby, 1963, p. 27–29, 46–50; Osberg, 1952, p. 87–90). The mineral lineation is best developed in quasi-flexural folds in the northwestern part of the eugeosynclinal zone; the slickensides and minor folds in flexural-slip folds are best developed in the miogeosynclinal zone and in passive-slip folds on some steep homoclinal limbs of major folds in the eugeosynclinal zone. Some of the mineral lineation, whose distribution and orientation is about the same as the distribution of the early cross folds in the northwestern part of the eugeosynclinal zone, is possibly of the early fold regime. Further detailed study is necessary to clarify this distinction. The flexural-slip folds occur especially in the vicinity of thrust faults down the dip of which they plunge (Balk, 1952; Thomas, 1968, p. 22–44); they are not to be confused either with the comparably oriented passive-slip folds, mentioned above, or with the early cross folds.[29]

The late folds are a part of the northwestward bulge of the New England salient, although they reflect this bulge less closely than do the geosynclines and geanticlines. They form anticlinoria that roughly coincide with the previously formed geanticlines, and synclinoria that coincide with the intervening and adjoining geosynclinal troughs (Pl. 3). The axial surfaces of subsidiary folds and related axial-plane cleavage are subparallel to the axial surfaces of the anticlinoria and synclinoria but dip less steeply. The axial surfaces of most late longitudinal folds in southern Vermont dip steeply southeast and the folds face northwest, but to the north in Vermont and in Quebec they are almost vertical (Béland, 1957, Map 1126; Doll and others, 1961; Osberg, 1965, p. 240–241). To the southeast, in eastern Vermont, New Hampshire, and neighboring areas, the late longitudinal folds are also nearly upright (Billings, 1955; Doll and others, 1961).

The trends and attitudes of folds bordering various tracts of late longitudinal folds vary with the shape and kind of adjoining tectonic features. As the domes and arches are approached, the orientation of the axial surfaces of the folds changes more or less gradually over distance of 1 to 10 miles (from that of the region) to subparallel to the flanks and tops of the domes and arches (Doll and others, 1961, sections *A-A′* to *F-F′*; White and Jahns, 1950, p. 209–213). Slip cleavage subparallel to the axial surfaces of the folds changes to a slip-cleavage schistosity on the domes (White, 1949). Where the folds encroach upon the craton they are tipped over toward the craton; moreover, the thrust faults dip east in the same direction as the axial surfaces of the folds (Clark, 1964a, p. 24; 1964b, p. 4–5; 1964c, p. 75–80; Doll and others, 1961, sections

A-A′ to *F-F′*; Dresser, 1912, p. 35; Lespérance, 1963, p. 6; Logan, 1862, p. 321; Osborne, 1956, p. 173, 193). These conditions are particularly apparent in Vermont, where the thrusts are most common and the pattern of folding is shown extensively by the contact between the Precambrian basement rocks and the overlying Paleozoic rocks, and also by the numerous relatively thin stratigraphic units characteristic of the lower Paleozoic portion of the miogeosynclinal zone.

The late folds of first magnitude in northwestern New England and adjacent Quebec are, from northwest to southeast, the Middlebury-Hinesburg-St. Albans synclinorium, the Green Mountain-Sutton Mountain anticlinorium, the Connecticut Valley-Gaspé synclinorium, the Bronson Hill-Boundary Mountain anticlinorium, and the Merrimack synclinorium. Only the northwestern part of the Merrimack synclinorium is included in the region of the present study.

MIDDLEBURY-HINESBURG-ST. ALBANS SYNCLINORIUM

Folds

The Middlebury-Hinesburg-St. Albans synclinorium (Cady, 1960, p. 535) is a north-trending tract that occupies much of the Champlain Valley in Vermont and immediately adjacent Quebec. It extends south of the Champlain Valley, chiefly beneath the Taconic Mountains in southwestern Vermont, western Massachusetts, and eastern New York. It is essentially a major foreland fold closely underlain by and attached to the Precambrian crystalline basement rocks at the southeast margin of the craton.[30] The synclinorium is a belt of transition from that of strong folding in the Green Mountain-Sutton Mountain anticlinorium to little or no folding on the craton. The total length of the synclinorium is more than 200 miles.

Two culminations divide the synclinorium into three parts: the Middlebury synclinorium (Cady, 1937) on the south, the Hinesburg synclinorium (Bain, 1931, p. 506–507), and the St. Albans synclinorium (Shaw, 1958, p. 563). (*See also* Cady, 1945, p. 562–564; Doll and others, 1961; Fisher and others, 1961.) The Middlebury-Hinesburg-St. Albans synclinorium contains lower Paleozoic parautochthonous miogeosynclinal and allochthonous eugeosynclinal rocks. It coincides fairly closely with a trough of miogeosynclinal subsidence between the Vermont-Quebec geanticline and the craton to the west and northwest, and appears to be a relic of this trough (Pl. 3).

The longitudinal folds of the Middlebury-Hinesburg-St. Albans synclinorium plunge gently in both directions from the two culminations. The St. Albans synclinorium to the north loses some of its clarity in the more uniform stratigraphic sequences of the eugeosynclinal zone and exogeosyncline southeast of the St. Lawrence River in Quebec (Cady, 1960, p. 540–541, Pl. 1). The Middlebury synclinorium[31] plunges south, includes the allochthonous eugeosynclinal rocks at the north end of the centrally located Taconic klippen, and is obscured in a more widespread terrane of both allochthonous and autochthonous pelitic rocks east of the Hudson River in the southern Taconic region, New York (Zen, 1963b). The axial surface of the synclinorium dips east and the axial surfaces of its subsidiary synclines and anticlines converge upward to the west, thus much of the east limb is overturned. Overturning is greatest in the pelitic terranes that characterize the uppermost levels, both stratigraphic and tectonic, of the synclinorium. In such terranes, because of the greater overturning, it is difficult to distinguish the late longitudinal folds from the early longitudinal and oblique folds.

Late minor folds closely reflect the gentle plunges of the major folds, and steeply plunging minor folds (such as are common in the longitudinal fold belts to the east of the Middlebury-Hinesburg-St. Albans synclinorium) are rare. In a few places where the early oblique folds are well developed, the late longitudinal folds appear to be partly or completely absent, apparently because the shallow rocks in which the early folds formed were less ductile during the late regime than the deeper rocks in which late folds did form (Crosby, 1963, p. 24–25, 41, 71–72; Zen, 1961, p. 324; *see also* Cady, 1945, Pl. 8, fig. 2, Pl. 9, fig. 4).

Thrust Faults

The principal thrust faults of the northwestern New England-Quebec region are east-dipping, essentially foreland thrusts, formed chiefly in the lower Paleozoic miogeosynclinal zone and genetically related to late longitudinal folding in the vicinity of the Middlebury-Hinesburg-St. Albans synclinorium.[32] Most of the thrusts are southwest of the axis of the New England salient; the largest ones, which extend south from this axis on both limbs of the synclinorium, die out in anticlines of the late fold regime, from which they developed by the breaking of anticlinal hinges (Cady, 1945, p. 565, 577). The thrust faults become difficult to trace northeast of the axis of the salient, where they pass from the miogeosynclinal carbonate-quartzite assemblage in the Champlain Valley into the eugeosynclinal and exogeosynclinal graywacke-shale assemblage southeast of the St. Lawrence River (Cady, 1960, Pl. 1). There, along Logan's Line, various geologic evidence indicates thrust or reverse faults, or both, that dip southeast and that are possibly offset by northwest-striking faults (Clark, 1951; 1964a, p. 24; 1964b, p. 4–5, 43; 1964c, p. 75–80; Osborne, 1956, p. 170–178, 196; Poole, 1967, p. 34–35).

The surfaces of the major thrusts dip gently (less than 20°) subparallel to either overlying or underlying strata, or to both. Competent strata in the overthrust rocks, where not previously deformed, commonly dip east parallel to the thrust surfaces. Incompetent strata fairly commonly found beneath the thrust surfaces are generally more disturbed. The surfaces of some small thrusts dip more steeply and are less concordant with adjoining strata.

The principal thrusts on and west of the west limb of the Middlebury-Hinesburg-St. Albans synclinorium are, from north to south, the Highgate Springs-St. Dominique, Philipsburg, Champlain, and Orwell thrusts (Cady, 1945, p. 565, 572; Clark, 1934, p. 6–8; Keith, 1923a, p. 104; McGerrigle, 1930, p. 181–186), of which the Champlain thrust is much the most extensive. (*See also* Doll and others, 1961; Welby, 1961, p. 193–199). Gently dipping Lower Cambrian dolomite and quartzite (Dunham Dolomite and Monkton Quartzite) of the west limb of the synclinorium overlie the Champlain thrust. Various Upper

Cambrian and Lower Ordovician dolomites, limestones, and quartzites overlie the other thrusts, all of which are truncated by the Champlain thrust. The principal footwall rocks are various Middle Ordovician shales that overlap westward on the Upper Cambrian and Lower and Middle Ordovician sandstones, dolomites, and limestones of the craton.

The principal thrusts along the east limb of the synclinorium are, from north to south, the Hinesburg thrust (Cady, 1960, p. 535, 539; Keith, 1932, p. 359, 364), the Monkton thrust (Cady, 1945, p. 574; Keith, 1932, p. 359, 364), the Pine Hill thrust (Brace, 1953, p. 78–79; Fowler, 1950, p. 62–63; Zen, 1964b, p. 56–57), the Dorset Mountain thrust (Thompson, 1959, p. 72), and the Maple Hill thrust (MacFadyen, 1956, p. 42–43), of which the Hinesburg thrust is the most extensive (*see also* Doll and others, 1961). Various Lower Cambrian units, most notably a folded massive quartzite (Cheshire Quartzite) more than 1000 feet thick, overlie the Hinesburg, Monkton, and Maple Hill thrusts. Precambrian basement rocks, as well as the Cambrian quartzite and an associated phyllite, overlie the Pine Hill thrust. Lower Ordovician limestone and dolomite overlie the Dorset Mountain thrust. The footwall rocks of the Hinesburg thrust are folded Cambrian and Lower and Middle Ordovician quartzite, dolomite, limestone, and slate units that are truncated by the thrust. The footwall of the Monkton thrust is Lower Cambrian quartzite (Monkton Quartzite). Truncation of folds in the footwall is less clear in the southern thrusts, although the Pine Hill and Maple Hill thrusts cut various Lower Cambrian to Middle Ordovician footwall rocks, and the Dorset Mountain thrust intersects one of the Taconic slides, between a Taconic klippe and underlying autochthonous Ordovician rocks. (*See also* Potter, 1963, Fig. 2, p. 61).[33]

Other thrusts also, most notably the thrust beneath the Sudbury nappe (Cady, 1945, p. 570; Crosby, 1963, p. 119–121), probably transect one of the Taconic slides. The thrust beneath the Sudbury nappe is eroded so that only its relations to the autochthon can now be seen; but before erosion, which separated the Taconic allochthon into two klippen at the north end of the Taconic Range, the Sudbury thrust probably extended up through the allochthon. Though the Sudbury thrust cuts the east limb of the Middlebury segment of the synclinorium, it partakes of the pattern of the south plunge of this segment, which indicates that thrusting preceded at least part of late longitudinal folding. The nappe is probably an early fold formed before the late folding and before the thrust.[34]

GREEN MOUNTAIN-SUTTON MOUNTAIN ANTICLINORIUM

Distribution

The Green Mountain-Sutton Mountain anticlinorium (Ells, 1887, p. 32–35; 1888, p. 89; Keith, 1923b, p. 318–324) forms a broad arc, convex to the northwest, that extends more than 400 miles northeastward from northwestern Massachusetts through Vermont and Quebec toward the northern tip of Maine. Its plunge is mostly northeast although it is abruptly south in Massachusetts. (*See also* Béland, 1962, p. 16; Cady, 1945, p. 564; 1960, p. 535; Doll and others, 1961; Dresser and Denis, 1944, p. 483; Lespérance and Greiner, in press; McGerrigle, 1934, p. 102, 106.) The anticlinorium is chiefly in lower Paleozoic eugeosynclinal strata, but miogeosynclinal rocks of this age occur on the west limb in western Vermont and on the east limb in southernmost Vermont. On the east limb, middle Paleozoic strata are miogeosynclinal. The axis of the Green Mountain-Sutton Mountain anticlinorium roughly coincides with that of the lower Paleozoic (Cambrian and Ordovician) Vermont-Quebec geanticline in the Green Mountains of central Vermont and in the Notre Dame Mountains in Quebec; but in the northern Green Mountains and in the Sutton Mountains in Quebec, the axis is southeast of the Vermont-Quebec geanticline (which there coincides with the Enosburg Falls anticlinal tract shown on Pl. 1) and it passes more directly northeastward across the New England salient than does the geanticline. In the southern Green Mountains it is west of the Vermont-Quebec geanticline, which swings southeastward toward the Berkshire Highlands in Massachusetts.[35]

It should be emphasized at this point that the Green Mountain-Sutton Mountain anticlinorium cannot possibly be the root of the eugeosynclinal rocks of the Taconic klippe, because it plunges southward and disappears in the southern Green Mountains in southern Vermont and northwestern Massachusetts where it encroaches upon the miogeosynclinal zone west of the Vermont-Quebec geanticline. It thus does not extend far enough south to have accommodated the full length of the Taconic klippe, nor does it contain rocks of appropriate facies near its south end. Hence the geanticline, which was already present before the folding of the anticlinorium and which extended southward almost entirely within the eugeosynclinal zone, is the most likely root of the klippe. (*See also* Rosenfeld, 1965).

The plunge of the longitudinal folds of the anticlinorium within the New England salient is mostly away from the axis of the salient (Cady and others, 1962; Cady and others, 1963, Pl. 1; Christman, 1959, Pl. 2; Christman and Secor, 1961, Pl. 2), and is opposite to the slopes of the sides of the transverse geosynclinal trough of the salient toward the axis.

The anticlinorial folds are weakest near the Vermont-Quebec geanticline, especially a little north of where the geanticline crosses the axis of the New England salient, which is near the international boundary.

Major Folds of the Green Mountain-Sutton Mountain Anticlinorium

The Green Mountain-Sutton Mountain anticlinorium includes a tract of various moderately persistent large doubly plunging axial anticlines in which lowest Paleozoic and Precambrian rocks are commonly exposed near the crests of the mountains. It also includes tracts of subsidiary doubly plunging anticlines and synclines that flank the axial anticlines to the northwest and southeast, particularly in the New England salient.[36] The axial surfaces of these folds diverge upward in the southern two-thirds of the salient, in northern Vermont and southern Quebec, dipping west in eastern tracts and east in western tracts (Doll and others, 1961, section *A-A′*, *B-B′*). In the northern third of the salient in Quebec, however, nearly all the axial surfaces dip northwest (Ambrose, 1942, section *A-B*; Osberg, 1965, Pl. 1, sections *B-B′* and *F-F′*; St. Julien, 1967, p. 46–47).

Axial Anticlinal Tract. The axial anticlines are most prominent in the more northeastern and southwestern parts of the anticlinorium, on either side of the New England salient and free of the flanking subsidiary anticlinal tracts. The Precambrian basement is exposed in the axial anticline in southern Vermont, whereas only lower Paleozoic rocks are exposed in the axial anticlines in northern Vermont and Quebec. In the New England salient, the axial anticlines are closely in right echelon—more so to the south than to the north (Albee, 1957; Cady and others, 1963, Pl. 1; Cady and others, 1962; Christman, 1959, Pl. 1; Christman and Secor, 1961, Pl. 1; Dennis, 1964, Fig. 8, de Romer, 1961, Fig. 7; Osberg, 1965, Fig. 15). The dextral pattern produced by the anticlines is in agreement with the dextral pattern of the minor folds (later to be discussed) that are found widely in the salient on the southeast limb of the anticlinorium.

The axial anticlinal tract is complicated by both cross folds and oblique folds of the early fold regime. Major cross folds, some of which are probably of the early regime but which veer into parallelism with the longitudinal folds of the axial anticlinal tract (Cady and others, 1963, p. 56–57, 63–67), are found near the axis of the New England salient. Major oblique folds of the early regime on both the southwestern and northeastern flanks of the New England salient (Osberg, 1952, Pl. 2, fig. 10; written commun., 1960, *cited in* Doll and others, 1961; Osberg, 1965, p. 241) are partly distinguishable from both early and late longitudinal folds parallel to which they swing. Where the oblique folds are most strongly developed the axial anticlines are subdued—at the southwestern flank of the New England salient only minor folds of the late regime are present and on the northeastern flank even the minor folds are difficult to distinguish. Both of these areas of oblique folds and subdued axial anticlines happen to be near the Vermont-Quebec geanticline where the axial anticlines are poorly expressed by the broadly flexed basement contact closely beneath.

Enosburg Falls Anticlinal Tract. The Enosburg Falls anticlinal tract in northwestern Vermont includes, in addition to the Enosburg Falls anticline (Doll and others, 1961), at least one other doubly plunging anticline northwest of the axial anticlines of the anticlinorium. This anticlinal tract is almost entirely within the New England salient in the foothills and lesser mountain ridges northwest of the Green and Sutton Mountains. As with the axial anticlinal tract, the overall plunge in all but the south end of the Enosburg Falls tract is to the north, away from exposures of the Precambrian basement rocks. The distribution of the principal anticlines is curvilinear rather than in echelon (Cady and others, 1962; Dennis, 1964, Fig. 8; Doll and others 1961; P.H. Osberg, written commun., 1960, *cited in* Doll and others, 1961; Obserg 1965, Fig. 15; Rickard, 1965, p. 525; Stone and Dennis, 1964, Fig. 5).

The Enosburg Falls anticlinal tract is complicated by major longitudinal and oblique folds of the early regime. The major early folds are least masked by the late longitudinal folds near and north of the axis of the New England salient, although the mutual parallelism of both sets has led to understandable confusion. Near the axis of the salient and coincident with the Vermont-Quebec geanticline, distinguishable late longitudinal folds are minor folds and the map pattern (Pinnacle Formation, including Tibbit Hill Volcanic Member, and Underhill Formation, including White Brook and Fairfield Pond Members of Doll and others, 1961) is determined by the major early longitudinal folds (Clark and Eakins, 1968, p. 168; Dennis, 1964, Pl. 1; Rickard, 1965, p. 524, 525,

Fig. 1, p. 526, Fig. 2A; Stone and Dennis, 1964, Pl. 1). The recognizable late longitudinal folds near the north flank of the New England salient in Quebec, are major folds and they transect obscure major early oblique folds near the St. Francis River, from which they are distinguished by their differing trends (Osberg, 1965, Fig. 15). Likewise, near the south flank of the salient in central Vermont, the major folds (shown by the areal pattern of the Precambrian basement) are of the late longitudinal set and the early folds are minor longitudinal folds that trend southward, then southeastward at the flank of the salient, into the large oblique folds of the early regime near the head of the White River (Osberg, 1952, p. 96–101).

Stoke Mountain Anticlinal Tract. The Stoke Mountain anticlinal tract (Ells, 1887, p. 31–32), the north end of which coincides with the Stoke Mountain geanticline, includes several doubly plunging subsidiary anticlines southeast of the axis of the Green Mountain-Sutton Mountain anticlinorium, within the New England salient. It is centered on lesser mountain ridges—namely, from southwest to northeast, the Northfield, Worcester, and Lowell Mountains in Vermont, and Bunker Hill and the Stoke Mountains in Quebec. The anticlines have linear to echelon pattern (Albee, 1957; Ambrose, 1943; Cady, 1956; Cady and others, 1963, Pl. 1; Cady and others, 1962; Duquette, 1959, p. 249; Ells, 1887, p. 31; Konig and Dennis, 1964, Pl. 1). This tract is possibly complicated by major longitudinal folds of the early regime that are not readily distinguished from the late longitudinal folds—except where the early folds are clearly truncated at unconformities that themselves have been folded (Cady and others, 1963, p. 26, Pl. 1; Lamarche, 1965, p. 176; 1967, p. 11; St. Julien, 1963a, p. 144; St. Julien and Lamarche, 1965, p. 13.)

Minor Folds of the Green Mountain-Sutton Mountain Anticlinorium

The minor folds of the Green Mountain-Sutton Mountain anticlinorium reflect the general orientation of the major anticlines and synclines (*see also* Hatch and others, 1967, p. 12), although they plunge more steeply in many places than do the major folds, and their surface trends are more varied. These minor folds are complex folds in that the axial-plane cleavage associated with them is commonly a slip cleavage whose sense of offset can be the opposite of the drag sense of the folds. This feature that is especially characteristic of synclinal tracts is also found in anticlinal tracts (Brace, 1953, p. 89; Hawkes, 1940, p. 142; Roth, 1965, p. 77).

The steeply plunging minor folds form two principal tracts, most apparent southwest of the axis of the New England salient (Fig. 3). One is the synclinal tract between the axial anticlines of the Green Mountain-Sutton Mountain anticlinorium and the Stoke Mountain anticlinal tract. The other is in southern parts of the homoclinal section southeast of the Stoke Mountain tract (southeast limb of the Worcester Mountain anticline—northwest limb of the Townshend-Brownington syncline, between the anticline and the Strafford-Willoughby arch). Within the south half of the New England salient, most of the minor folds on the southeast limb of the anticlinorium, including those with anomalously steep plunges, are south-plunging dextral folds (Albee, 1957; Cady, 1956; Cady and

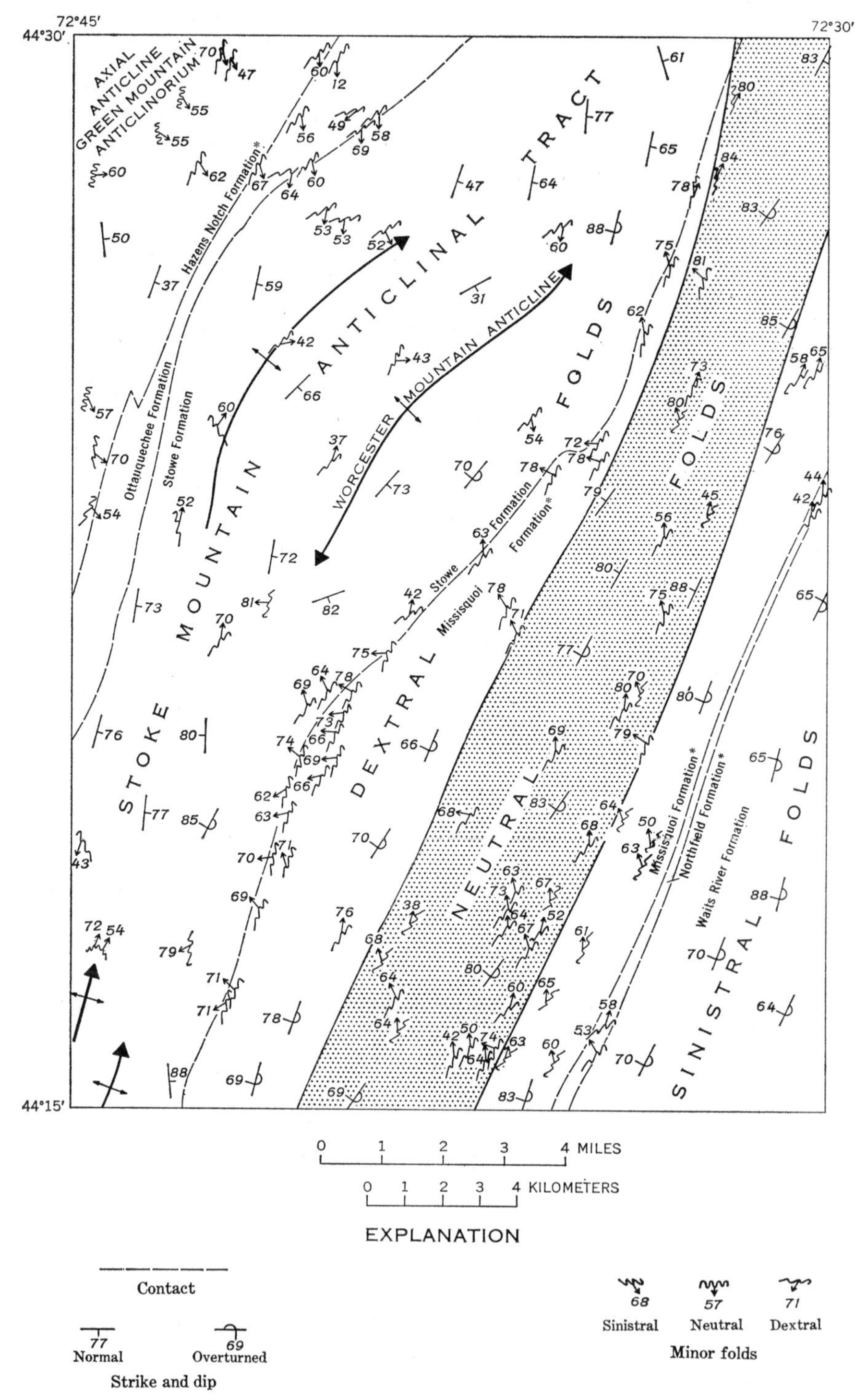

Figure 3. Tectonic map of the Montpelier quadrangle, Vermont, showing belts of steeply plunging dextral, neutral, and sinistral folds (outlined on Pl. 1.)
* Doll and others, 1961.

others, 1962; Christman, 1959, Pl. 1; Christman and Secor, 1961, Pl. 1). Thus their plunge and movement sense are such as would be expected on the southeast limb of a southwest-plunging anticlinorium, though the sense of offset on axial-plane slip cleavage is commonly the opposite. (*See* Fig. 4.)

The belt of steeply south plunging dextral folds extends a little into the north half of the New England salient, on the east limbs especially of the axial anticlines of the Green Mountain-Sutton Mountain anticlinorium in Vermont (Cady and others, 1963, Pl. 1). Near the international boundary, however, the southward plunges of the dextral folds become more gentle, and yet farther north in Quebec as the northeast flank of the salient is approached, these folds pass into gently northeast-plunging sinistral folds, expectable on the east limb of the northeast-plunging axial anticline of the anticlinorium (de Römer, 1961, Fig. 7; Osberg, 1965, Pl. 1, section *F-F′*; St. Julien and Lamarche, 1965, p. 13). Thus the belt of dextral folds tapers northward and disappears near the international boundary.

CONNECTICUT VALLEY-GASPÉ SYNCLINORIUM

Distribution and Relation to the Domes and Arches

The Connecticut Valley-Gaspé synclinorium (Cady, 1960, p. 536; Duquette, 1959, p. 246; Marleau, 1959, p. 129) occupies a long curvilinear tract, chiefly on the west side of the Connecticut River in western New England and southeast of the Notre Dame and Shickshock Mountains respectively in southeastern Quebec and northernmost New England (Pl. 1), and in the Gaspé Peninsula, Quebec (Fig. 1). This tract extends north and northeast in an arc of 750 miles, from Long Island Sound to the Gulf of St. Lawrence (Fig. 1). (*See also* Doll and others, 1961.) It contains middle Paleozoic strata transitional between those of the miogeosynclinal and eugeosynclinal zones, in northwestern New England and adjacent Quebec. It rudely coincides with a geosynclinal trough between the Stoke Mountain and Somerset geanticlines in northern New England and Quebec and southeast of the southern part of the Vermont-Quebec geanticline.

The axial surface of the synclinorium is nearly upright in northern Vermont and nearby Quebec, but in southern Vermont and Massachusetts it is tipped over to the west so that it dips east (Doll and others, 1961). The axis plunges gently northeast in agreement with the overall plunge of the Green Mountain-Sutton Mountain anticlinorium. Near the northwest margin of the synclinorium, a few fairly well defined major folds plunge gently in the same direction (Doll, 1951, p. 50, 52; Konig and Dennis, 1964, Pl. 1). However, the configuration of most of the folds in the synclinorium is determined principally by the pattern of the domes and arches, already discussed, that occupy large areas in the vicinity of the synclinorial axis, especially where it crosses the New England salient. The axial surfaces of the folds converge upward over the domes, toward and more or less symmetrically with the axial surface of the synclinorium. Thus the axial surfaces of normal drag folds that face up the east and west limbs of the synclinorium are subparallel to the axial surfaces of reverse drag folds that face down the east and west flanks of the domes and arches (Fig. 4A). Steeply northeast plunging sinistral folds on the northwest limb of the synclinorium become

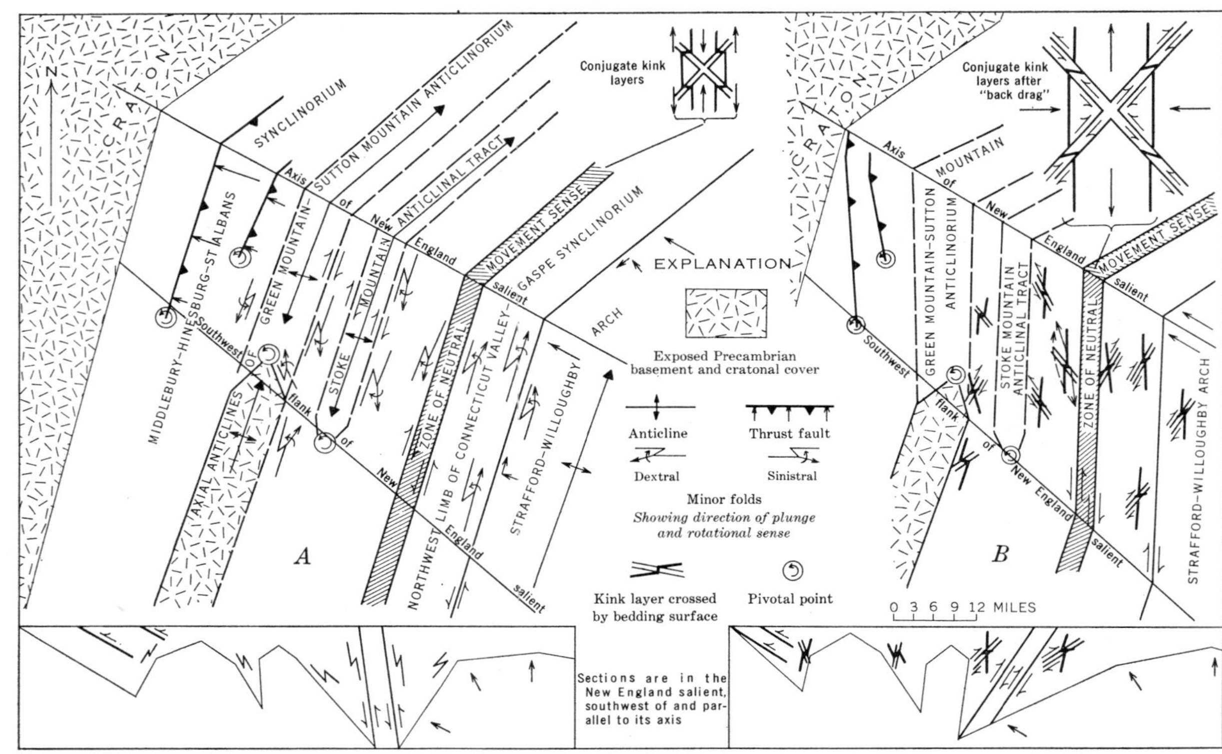

Figure 4. Mechanical diagram modeling doming and arching and late fold and thrust movement southwest of the axis of the New England salient, in northern Vermont. A. Plan and cross section of flexural folds at the beginning of thrusting; plan of related conjugate kink layers showing neutral movement sense. B. Plan and cross section of partly passive (quasi-flexural) folds formed by "back-drag" on slip cleavage bounding kink layers, possibly after thrusting; plan of related conjugate kink layers showing flattening.

more gently plunging sinistral folds on the northwest flank of an arch. The effect of this relationship is that the normal drag folds of the limbs of the synclinorium merge imperceptibly with the "reverse" drag folds of the flanks of domes and arches, across the axial surfaces of nearly isoclinal synclines (not shown in Fig. 4), and the movement senses of the normal and reverse drag folds are identical.

Major Folds of the Connecticut Valley-Gaspé Synclinorium

The two principal folds of the synclinorium are the Brattleboro-Averill syncline and the Townshend-Brownington syncline, respectively east and west of the belt of domes and arches in Vermont (Pl. 1; Doll and others, 1961). In some places these late folds possibly coincide with the reversed limbs of recumbent anticlines of the early regime (Ern, 1963, p. 60–72; Goodwin, 1962; 1963, p. 98–101; Woodland, 1965, p. 115–117), so they may be provisionally considered synforms in the regional foliation. They are gigantic similar (passive) folds, with passive axial surfaces as well as limbs, and their axes coincide with the lowest points in a given regional foliation surface. This surface is flexed in the various closely folded late synclines between the approximately longitudinal folds on the limbs of the Connecticut Valley-Gaspé synclinorium, and in the variously oriented folds on the flanks of the domes and arches. The Townshend-Brownington syncline, shown in southern Vermont by the pattern of laterally persistent stratigraphic units from Precambrian to middle Paleozoic, is traced more easily than the Brattleboro-Averill syncline. These synclines extend elsewhere into zones of both lateral facies change and early recumbent folding, where because of such complexities the map pattern alone is an unreliable clue to the stratigraphic succession and to the distribution of the late folds. (*See* Ern, 1963, p. 71–72; Konig, 1961, p. 69–74; Murthy, 1957, p. 22–24, 44–46; 1958).

The late longitudinal folds out on the limbs of the Connecticut Valley-Gaspé synclinorium merge imperceptibly with those on the limbs of the Green Mountain-Sutton Mountain anticlinorium to the northwest and of the Bronson Hill-Boundary Mountain anticlinorium to the southeast. Especially in the New England salient, the transitions from the synclinorium into the adjoining anticlinoria are steep homoclines, the traces of the contacts between lower and middle Paleozoic units are broadly curvilinear (Doll and others, 1961; Quebec Department of Mines, 1957) and most folds are minor.

Minor Folds of the Connecticut Valley-Gaspé Synclinorium

The minor folds of the Connecticut Valley-Gaspé synclinorium merging with those of the domes and arches, plunge gently northeast in the same direction as the synclinorium. The folds on the northwest limb are sinistral (Cady, 1950a, Pl. 1, fig. 1; 1956; Ern, 1963, p. 60, 68, Pl. 2; Jahns, 1967, Fig. 5. 3; Konig, 1961, p. 64, 75, Pl. 2; Konig and Dennis, 1964, p. 48; White, 1959, p. 579–580; White and Jahns, 1950, p. 203–206) and those on the southeast limb are dextral (Eric and Dennis, 1958, p. 35–38, 41, Pl. 2; Hatch, 1963, Fig. 15, Pl. 2; Johansson, 1963, p. 56–57, Pl. 2; Lyons, 1955, p. 131, Pl. 2; White, 1959, p. 579–580; White and Billings, 1951, p. 679, 680, 681, Pl. 3; White and Jahns, 1950, p. 208;

see also Cady, 1960, p. 552). These plunges reflect the general northeasterly regional plunge more faithfully than do the minor folds of the Green Mountain-Sutton Mountain anticlinorium, which are shown to plunge to the southwest in at least the south half of the New England salient.

Though the late minor folds on the limbs of the Connecticut Valley-Gaspé synclinorium commonly plunge gently subparallel to the direction of strike there are places, especially on the northwest limb of the synclinorium, where they plunge very steeply, approximating the direction of dip of the foliation (Fig. 3). In these places, the fold axes appear to vary in orientation nearly through the vertical (Cady, 1956; White and Jahns, 1950, p. 203–204) instead of through the horizontal. The sinistral pattern of the folds that plunge steeply northeast on the northwest limb of the synclinorium is expectable on this limb. This pattern is, moreover, expectable on the northwest flank of a north-plunging arch.

This belt of steeply plunging sinistral folds, on the northwest limb of the Connecticut Valley-Gaspé synclinorium, is parallel to and adjoins the belt of steeply plunging late minor dextral folds, previously discussed, on the southeast limb of the Green Mountain-Sutton Mountain anticlinorium (Fig. 4). The change between these belts takes place a little northwest of the contact between the lower and middle Paleozoic rocks (Middle Ordovician Moretown Member of the Missisquoi Formation and Silurian Shaw Mountain Formation—Cady, 1956) that is used arbitrarily as the boundary between the synclinorium and the anticlinorium, in northern Vermont (Fig. 3). In that area the folds plunge steeply north, but several miles to the northwest in the Stoke Mountain anticlinal tract (Worcester and Northfield Mountain anticlines), the plunges change through the vertical to steeply south.

The relative quantities of the sinistral and dextral folds vary within the zone of transition from the belt of sinistral folds to that of dextral folds. Some of the folds form nearly vertical sets of conjugate kink layers in which the sets containing dextral folds strike northwest and those with sinistral folds strike northeast (Fig. 4A). The conjugate kink layers especially suggest that the transition is very subtle. The bulk of the rock moved neutrally (sinistral and dextral folds being about equal in number), and structural continuity was maintained between the rocks to the southeast that contained chiefly sinistral folds, and those to the northwest that contained chiefly dextral folds. During formation of the flexural folds, the rocks in this neutral zone apparently moved south and down relative to the belt of north-plunging sinistral folds to the east, and up relative to the belt of south-plunging dextral folds to the west. It should be remembered, however, that the kink layers commonly show at their bounding surfaces, offsets of bedding and bedding schistosity whose sense is the opposite of that of the kinks (Fig. 4B). Hence the neutral zone of the conjugate kink layers was flattened parallel to the regional foliation—after the layers were formed in connection with flexural folding. This is to say that with continued deformation the style of flexural folds approached that of passive folds (quasi-flexural folds), and the rocks of the neutral zone eventually moved north and also up relative to the belt of sinistral folds and down relative to the belt of dextral folds.

BRONSON HILL-BOUNDARY MOUNTAIN ANTICLINORIUM

Distribution

The Bronson Hill-Boundary Mountain anticlinorium (Billings, 1937, p. 519; Cady, 1960, p. 536) forms a curvilinear tract convex to the northwest that rudely coincides with the north-south course of the Connecticut River in western New England and extends northeastward along the Boundary Mountains between Quebec, northern New Hampshire, and northwestern Maine for a total distance of at least 250 miles (Fig. 1). (*See also* Albee, 1961; Billings, 1955; 1956, p. 109–110; Doll and others, 1961; Green, 1964.) It plunges generally to the south but fairly abruptly plunges northeast in northwestern Maine. Eugeosynclinal lower and middle Paleozoic rocks are exposed in the anticlinorium; the middle Paleozoic contains thin units (Silurian Clough and Fitch Formations) of carbonate-quartzite assemblage as in the miogeosynclinal zone, but is underlain and overlain by eugeosynclinal strata (Cady, 1960, p. 544–545). The northern part of the Bronson Hill-Boundary Mountain anticlinorium coincides fairly closely with the Somerset geanticline.

The longitudinal anticlinorial folds in the New England salient plunge chiefly toward the salient axis; this contrasts with plunge away from the axis of the salient in the Green Mountain-Sutton Mountain anticlinorium. The change from northeast plunge (Hadley, 1942, p. 154, 156–157; 1950, p. 28–29) to southwest plunge (Eric and Dennis, 1958, p. 29–30; Johansson, 1963, p. 59–60) occurs in a distance of less than 20 miles from northeast to southwest in which the axes of the folds, most commonly synclines, are nearly horizontal or doubly plunging (Billings, 1937, p. 519–524; White and Billings, 1951, p. 670–673). The pattern of the southeasternmost fold of the anticlinorium (Bronson Hill anticline) shows the change of plunge especially well (Billings, 1955). This change in plunge is remarkably close to a fairly well marked change in trend of the Bronson Hill-Boundary Mountain anticlinorium where it crosses the axis of the salient. Within 40 miles northeast of the axis of the salient, well out on its flank, plunges are both to the northeast and southwest (Green, 1968, p. 1628–1632; Green and Guidotti, 1968, p. 263; Hatch, 1963, p. 52–53, 55), and the effects of the salient, if any, are not clear.

Major Folds of the Bronson Hill-Boundary Mountain Anticlinorium

The Bronson Hill-Boundary Mountain anticlinorium is made up of doubly plunging major folds. These folds are partly like those of the Green Mountain-Sutton Mountain anticlinorium, but only to the northeast are the anticlinal tracts discrete and fairly continuous. The axial surfaces of the folds diverge upward to some extent, but the anticlinorium as a whole appears to be nearly upright. (*See* structure sections in Billings, 1955, and in Doll and others, 1961.)

Two contrasting styles of major folds are represented. One is typified by anticlines similar to those of the Green Mountain-Sutton Mountain anticlinorium that have subsidiary folds of all sizes. The other is characterized by broad arched and fairly commonly domal anticlines ("Oliverian" domes,

Billings, 1955) that have few subsidiary folds. The differences in style appear to depend chiefly upon the kind of rock exposed in the axial parts of the anticlines. Interbedded granofels, feldspathic quartzites and phyllites are exposed in the axial areas of the anticlines with subsidiary folds. Fairly massive and mainly bedded granitic rocks are in the domal anticlines. (*See also* Billings, 1937, p. 535–536; Naylor, 1968, p. 234–238; Thompson and others, 1968, p. 214–215.) The anticlines and synclines with subsidiary folds are discussed first.

The Gardner Mountain anticline, which is the northwesternmost major fold of the anticlinorium (Billings, 1937, p. 519–520; Eric and Dennis, 1958, p. 29–30; White and Billings, 1951, p. 671–672), is the largest in the axial region of the New England salient. This anticline possibly continues into other less extensive anticlines to the northeast (Hatch, 1963, p. 55) and southwest (Hadley, 1942, p. 154; 1950, p. 28–29), somewhat beyond the flanks of the salient.

Other anticlines and synclines southeast of the Gardner Mountain anticline are smaller and less continuous along the strike (Billings, 1937, p. 520–524; Eric and Dennis, 1958, p. 29; Hadley, 1942, p. 156–158; Johansson, 1963, p. 59–63; White and Billings, 1951, p. 670–671). The Lunenburg anticline, which is probably the most continuous to the northeast, is in line of trend with the axial anticline of the northeast end of the anticlinorium and appears to be continuous into it (Doll and others, 1961; Hatch, 1963, Pl. 3). In this area, the anticlinorium is properly known as the Boundary Mountain anticlinorium (Albee, 1961). The Salmon Hole Brook syncline persists farthest to the southwest, but becomes lost among the domal anticlines about to be discussed.

The domal anticlines, referred to collectively as the Bronson Hill anticline (Billings, 1955, 1956, p. 109–110), lie in a curvilinear tract at the southeast side of the Bronson Hill-Boundary Mountain anticlinorium. They are 15 to 20 miles southeast of the axis of the Boundary Mountain anticlinorium proper in northern New Hampshire. To the southwest, near the axis of the New England salient, this tract converges to within 10 miles of the axis of the Boundary Mountain anticlinorium and at the southern boundary of New Hampshire it nearly coincides with the axis. The domal anticlines are much more extensive to the south along the Bronson Hill-Boundary Mountain anticlinorium in Massachusetts and Connecticut (Brookins and Hurley, 1965, p. 3–4; Eaton and Rosenfeld, 1960, p. 169, 176; Lundgren, 1962, p. 5, Fig. 1, p. 15–16; Rodgers and others, 1959, p. 19–24). A few domal folds are west of the belt of their principal occurrence. These include a dome north of Lebanon in New Hampshire (Lyons, 1955, p. 127–130) and several that are actually on the west limb of the Connecticut Valley-Gaspé synclinorium in Massachusetts and Connecticut (Stanley, 1964, Fig. 2, p. 7—domes E, F, G, H, I, J).

Minor Folds of the Bronson Hill-Boundary Mountain Anticlinorium

The plunge of the minor folds reflects that of the major folds, but as elsewhere, some steep plunges closer to the dip of the foliation than to the strike indicate strike movement disproportionately larger than suggested by the overall gentle plunges of the major folds. The minor folds on the southern part of the west limb of the Salmon Hole Brook syncline plunge steeply southwest although the sinistral plan of stratigraphic units which top to the east, suggests a northeast plunge. Evidently the eastern rocks moved north relative to the western enough to tilt the fold axes through the near vertical (Billings, 1937, p. 522–523). Similarly oriented sinistral minor folds between two of the domal anticlines have been explained as caused by opposed horizontal components of movement during their growth (Lyons, 1955, p. 129, 130). (*See also* Stanley, 1967, p. 54–55, 57, 60.)

The late minor folds are in some places isoclinal and show well-developed axial-plane schistosity. Thus they closely resemble minor folds of the early regimes. However, bedding schistosity is preserved locally at the noses of the folds (Heald, 1950, p. 70), indicating that they are probably late.

MERRIMACK SYNCLINORIUM

The Merrimack synclinorium (Billings, 1955; 1956, p. 114) occupies a wide tract in south-central New Hampshire southeast of the Bronson Hill-Boundary Mountain anticlinorium that extends southward into Massachusetts and Connecticut and possibly northeastward into Maine. It is for the most part southeast of the region of this synthesis. It is a relic of a geosynclinal trough southeast of the Somerset geanticline. It probably contains both lower and middle Paleozoic rocks (Billings, 1956, p. 34–35; Thompson, 1954, p. 38), although to date, only one stratigraphic unit (Devonian and Silurian? Littleton Formation) has been mapped over much of the area included. It is made up of nearly upright folds of varied dimensions that trend north-northeast (Osberg and others, 1968, p. 245).

INTERPRETATION OF THE LATE FOLDS AND THRUST FAULTS

The late folds and thrust faults, like the early folds and slides, probably reflect (though indirectly) subhorizontal movements in response to gravity. The stresses involved were transmitted farther than in the early fold regime because of greater competency of the stressed rocks. Thus, the competent Precambrian basement was flexed and its cross section shortened as much as 40 percent—visibly in the vicinity of the Green Mountain-Sutton Mountain anticlinorium—and it participated probably in as much as 10 miles of overthrusting. Competent Paleozoic interbeds controlled the shape of lesser folds, and competent Paleozoic stratigraphic units that overlie some of the thrust faults transmitted some of the stress that accounted for thrusting. Insofar as the basement was clearly involved in the major folds of the late regime, the geanticlines (Pl. 3) that formed in the early Paleozoic probably moved closer together during the Devonian Acadian orogeny.

The late folds perhaps formed in response to outward spread from the rising domes and arches, with whose flanking folds they blend (Fig. 4A). Also, where the domes and arches plunge northeastward in Quebec, the folds are smaller. The principal spread, as with the early folds and slides, was westward and northwestward toward the craton, which most of the folds and overthrust slices face. To the southeast, where the folds are more nearly upright, horizontal movement appears to have been less. Where domes and arches predominate, as in the belt of the Connecticut Valley-Gaspé synclinorium—also the site of the geosynclinal trough between the Somerset geanticline and the Stoke Mountain and Vermont-Quebec geanticlines (Pl. 3)—it seems at first consideration less likely that the early Paleozoic geanticlines moved closer together during the Acadian orogeny. This hypothesis, of course, requires the assumption that material was added to the domes and arches from depth entirely by simple upwelling. If, however, (as seems probable) rock was also squeezed up from the Precambrian basement, the doming and arching may be considered to be part of the phenomenon of transverse shortening under lateral compression. (*See* Appendix A.) It is assumed here that squeezing as well as buoyant rise took place and that Acadian shortening was appreciable (Pl. 3), although it may not have reached 40 percent as in the vicinity of the Vermont-Quebec geanticline.

The southeast opening "bay" of the New England salient provided a framework into which "waves" of rock movement from the southeast were deflected. Thus, the movement was more free near the axis of the salient. In that area, the least competent Paleozoic rocks were thickest, and they piled up to the northwest against the "shallows" of the craton as folds that plunge northeast and southwest toward the margins of the salient (Fig. 4A). The thick eugeosynclinal section in the salient was also freer from basement control of deformation, so that steeply plunging minor folds developed independently of the major longitudinal folds that have Precambrian basement rocks at their core. Southeast of the domes and arches, in which direction the salient broadens and the rocks are less constricted, the slopes of the sides of the original trough of the salient apparently determined the plunge of the late folds toward the axis of the salient.

The responsibility of the New England salient for horizontal deflections of rock movement is illustrated by an interpretation of folding and thrusting in the Middlebury-Hinesburg-St. Albans synclinorium (as shown by Plate 1 and Figure 4).

The original trend of the synclinorium, when the late longitudinal folds started to form, was northeast, near the southeastern margin of the craton. The thrust slices developed from overturned and broken folds. Thus the major thrusts, which bound the two sides of the synclinorial tract, appeared beneath the competent strutlike gently dipping west limb and beneath the overturned east limb (west limb of the Green Mountain-Sutton Mountain anticlinorium). Thrusting was preceded by folding both of the competent sandstones, dolomites, and limestones in the miogeosynclinal zone and of the less competent eugeosynclinal shales and graywackes to the north in the northern part of the St. Albans segment of the synclinorium. During the folding the northwestern front of the competent rocks, less deformed within themselves (especially on the gently dipping west limb of the synclinorium) tended to move ahead of the less competent rocks to the north and to override the craton. Thrusting probably started with tear faulting along the boundary between the miogeosynclinal and eugeosynclinal facies at the north end of the present thrust slices in Quebec. As deformation continued, the thrusts extended by ripping southward along the hinges of the pre-existing anticlines, which were oriented in echelon with respect to the cratonal margin. (*See* Hubbert, 1928, p. 83–84.) The ripping stopped in the resistant Precambrian basement on the southwest flank of the New England salient, but the related counterclockwise movement of the slices and included folds continued. This is shown by the northward trend of the trace of the Champlain thrust athwart the traces of the Philipsburg, Highgate Springs-St. Dominique, and Orwell thrusts, which are apparently pivoted somewhere beneath the Champlain thrust slice in the northern Champlain Valley-St. Lawrence Valley region. (*See* Cady 1945, p. 575, 577–578; Keith, 1933, p. 52.)

South of the New England salient, pivotal movement of thrust slices and truncation of footwall folds are not apparent and the thrust surfaces themselves are warped into partial parallelism with the folded bedded rocks of the synclinorium. This is perhaps because without the pivotal movement, possible only in the unobstructed salient, continued stress could be relieved only by folding.

The steeply plunging minor folds, most abundant in the New England salient southeast of the axial anticlines of the Green Mountain-Sutton Mountain anticlinorium (Fig. 3), illustrate the complicated vertical and horizontal rock movements characteristic especially of the deep eugeosynclinal zone during late folding. The steep plunges were probably caused by deformation that continued after the major folds had become established. (*See* Weiss, 1958, p. 34.) Late minor folds and related bedding-cleavage intersections that plunge steeply downdip on the limbs of the major folds are an example. They probably formed in response to differential horizontal movements between adjacent major folds and between the major folds and the dome and arches (Fig. 4). The movements were principally in the direction of strike and probably caused both the steepening of the axes of earlier formed folds and the appearance of new steeply plunging folds. (*See also* Green, 1968, p. 1631.)

Steeply plunging minor dextral folds in the homoclinal section southwest of the axis of the New England salient and on the east side of the south plunging Stoke Mountain anticlinal tract, show that rocks to the east moved south relative to those to the west. High strata moved up and to the west and a considerable distance to the south over low strata. This is best explained by concomitant growth of the Green Mountain-Sutton Mountain anticlinorium and counterclockwise turning of both the axial and Stoke Mountain anticlinal tracts around well-anchored pivotal points on the southwest flank of the New England salient (Fig. 4A). Thus the anticlinal tracts, stiffened by metamorphic rocks (including the Precambrian basement raised in their cores) and cross braced by the broad span of the bedded rocks in their axial regions, were active structural units. The synclinal tracts and their homoclinal limbs, on the other hand, remained passive. The pivotal points are on the southwest flank of the salient (in central Vermont, near the head of the White River, Pl. 1), where the late minor folds and crenulations on the east limb of the anticlinorium change direction of plunge from south (Cady and others, 1962) to north (Osberg, 1952, Pl. 2). In that area, the rocks on the east limb of the anticlinorium moved upward and to the west and also to the north around the north-plunging surface of the Precambrian basement (in the axial anticline in the southern Green Mountains, Fig. 4A). (*See also* Hawkes, 1941, p. 658; Brace, 1953, Pl. 3.)

The counterclockwise movement of the belt of steeply plunging dextral folds southwest of the axis of the New England salient on the east limb of the Green

Mountain-Sutton Mountain anticlinorium thus compares with that of the thrust slices in the same relative position in the salient, but to the west of the anticlinorium (Fig. 4A). These movements are believed to have been parts of a general counterclockwise movement of the whole south side of the salient, most unobstructedly to the northwest near the axes of the salient. (*See* Hubbert, 1928, p. 83–84.)

The steeply plunging minor sinistral folds west of the domes and arches on the northwest limb of the Connecticut Valley-Gaspé synclinorium (Fig. 4A), bespeak movement in which rocks to the east moved north as well as up, relative to those to the west. This movement, which is otherwise expectable on the northwest limb of the northeast-plunging synclinorium, was probably accentuated by upward and northwestward expansion into the New England salient of a southwestern dome of the Strafford-Willoughby arch. The sinistral folds, like the dextral folds discussed in the preceding paragraphs, are southwest of the axis of the salient, and their distribution is probably determined by the relationship of the Strafford-Willoughby arch to the salient. The long axis of this arch, as already described, is symmetrically bisected by the axis of the salient. And the sinistral folds are in a belt of differential movement between the southwest quarter of the arch and the southwest flank of the salient into which the arch expanded. Movements are not as clear northeast of the axis of the New England salient, where those of opposite sense are expectable.

A zone of neutral-movement sense, between the belts of dextral and sinistral drag folds (Figs. 3, 4A), shows large components of both apparent and real southward movement, southwest of both the axis of the New England salient and the Strafford-Willoughby arch. The apparent southward movement contrasts with northward and upward movement within the west flank of the arch; the real movement is to the south and upward on the east limb of a south-plunging anticline of the Green Mountain-Sutton Mountain anticlinorium. Thus, if we assume that rise and expansion of the Strafford-Willoughby arch into the New England salient was the cause of folding in the vicinity, we can assume that the rocks in the belt east of the neutral zone moved upward and northwestward. This movement would have caused the belt of the neutral zone to be rotated counterclockwise, and where it pressed against the adjacent belt of the Green Mountain-Sutton Mountain anticlinorium to the west, the anticlinorium and the adjoining Middlebury-Hinesburg-St. Albans synclinorium would have also been rotated counterclockwise.

The partly passive folding (quasi-flexural folding) that followed flexural folding involved movements quite the opposite of those of the flexural folds (Fig. 4B). In the neutral zone, the slip-and-flow of passive deformation was to the north—also up, relative to the earlier belt of left-lateral flexural fold movement in the arch to the east—down, relative to that of right-lateral flexural fold movement in the anticlinorium to the west. This slip-and-flow is interpreted as flattening of the previously formed flexural folds (Fig. 4A), which, as pointed out in the description of conjugate kink layers, was clearly what happened within the neutral zone. The flattening probably occurred when, because of continued arching and possibly because thrusting was blocked by cratonal barriers on the south flank of the New England salient, the plastic environment of passive folding was approached. As it flattened, the rock moved north toward the axial zone of the New England salient and also updip to the east, west of the Strafford-Willoughby arch. The west-dipping, east-topping strata in the belt west of the Strafford-Willoughby arch were probably tipped over to the east by westward spread of the deep parts of the expanding arch. Components of this deep spread probably also tipped the axes of the once steeply south plunging flexural folds through the vertical to a steep northerly plunge on the east limb of the Green Mountain-Sutton Mountain anticlinorium in the vicinity of the neutral zone (Figs. 3, 4*B*). Thus the movements that included passive folding produced S-tectonites and were entirely different from those of the earlier flexural folding that produced B-tectonites (*see* footnote 28), and that culminated in thrusting and counterclockwise sweep of the strutlike anticlinal tracts of the Green Mountain-Sutton Mountain anticlinorium.

FOOTNOTES

[26] The parallel or concentric folds of Van Hise include approximately Donath's (1963) flexural folds, which form under conditions of relatively low mean ductility. (*See also* Carey, 1962, p. 97–98, 116–120; Rast, 1964, p. 178–179.)

[27] The kink layers have been variously referred to as "fold bands" (Cady, 1956) and "fold layers" (Cady and others, 1963, p. 61–62), but are what are fairly commonly also known as "kinkbands." They are here called "kink layers" out of deference to their three-dimensional aspect, which is very clear in these rocks. They grade into widely spaced coarse crenulation cleavage, where they are separated at their bounding surfaces from adjacent rock.

[28] The drag folds and related fracture cleavage, crenulation cleavage, and simple kink layers, without evidence of offset parallel to the planar elements, are features of B-tectonites; whereas an S-tectonite is likely especially where the offset is opposite in sense to the drag folds. B- and S-tectonites have been briefly discussed by de Sitter (1964, p. 87–88) and by Turner (1948, p. 199–200), and in detail by Phillips (1937, p. 587–599).

[29] Early and late cross folds occur together and have been distinguished at one place in the eugeosynclinal zone. This is near the axis of the Green Mountain-Sutton Mountain anticlinorium in the Sterling Pond area, north-central Vermont (Chidester, 1953, Table 1). In that area, late cross folds are fairly large culminations in the axial anticline of the anticlinorium that are considered to have formed at about the same time as the axial anticline.

[30] *Foreland* as used here refers only to the mobile shelf, which is marginal to the craton.

[31] The Middlebury synclinorium has recently been identified with the early fold regime because it was found difficult to differentiate the early oblique folds (Sudbury nappe) from the folds of the late regime represented on the east limb of the synclinorium (Voight, 1965). (*See* footnote 34, p. 80.)

[32] Some major thrust faults that have been mapped in belts southeast of the foreland are not included in the present discussion, either because they transect the longitudinal folds indiscriminately and appear to be genetically unrelated to the folds (Billings, 1933, p. 148–150; 1937, p. 525–530) and are possibly gravity (high angle) faults, or because they are more readily explainable as unconformable overlap (Billings, 1956, p. 97–98) or by sedimentary and volcanic facies variations (Cady, 1960, p. 552). Those that are clearly faults are discussed in a later section.

[33] Zen (1967, p. 32–33) suggests that it is impossible to determine whether the Dorset Mountain thrust formed before, at the time of, or much later than the Taconic slide included in its hanging-wall block, but concludes without presenting further evidence, "If the second choice is correct, then the carbonate and high Taconic rocks [in the hanging-wall block] moved as a single unit during the second episode." It should be noted at this point that Thompson (1959, p. 72) had concluded that "the Dorset Mountain thrust offsets the main Taconic fault [Taconic slide]," whereas Zen (p. 32) observes only that the Dorset Mountain thrust "is clearly later than emplacement" of rocks of the Taconic allochthon next west. Zen apparently assumes that the latter rocks make up a different "low Taconic" allochthonous slice ("Bird Mountain slice")

over which a more eastern allochthonous slice ("Dorset Mountain slice")—preserved above the Taconic slide in the upper part of the parautochthonous hanging-wall block of the Dorset Mountain thrust—slid westward before the Dorset Mountain thrust formed. These complexities reduce to the problem of whether the surface of dislocation west and southwest of Dorset Mountain is the Dorset Mountain thrust beneath parautochthonous rocks, as distinguished by Doll and others (1961, solid sawteeth on upper plate), or as Zen prefers, a Taconic slide beneath the Taconic allochthon and probably nearly coincident in distribution and time of formation with the Dorset Mountain thrust. If the surface of dislocation is a slide beneath completely allochthonous rocks, then as described by Zen (1967, p. 14-28), it separates a "high Taconic sequence" in Dorset Mountain and south from a "low Taconic sequence" to the west. On the other hand, if the surface is a thrust fault beneath parautochthonous rocks then the "high" and "low" Taconic rocks east and west of its trace are parts of a single allochthonous slice. The latter interpretation seems most likely inasmuch as the lithic differences between the allochthonous rocks to the east and those to the west are not striking (Doll and others, 1961). Moreover, stratigraphic units within the allochthon to the west as well as within the parautochthon and its contained allochthon to the east in Dorset Mountain are truncated by the thrust (Thompson, 1959, p. 86).

[34] The conclusion that the Sudbury thrust probably formed later than the Sudbury nappe conflicts with the author's original interpretation (Cady, 1945, p. 570), which is nearly the same as that of Voight (1965), who refers to the thrust as "a surface of discontinuous flow separating the nappe from the core of the subjacent [Middlebury] synclinorium" and "contemporaneous with the nappe." The early northwest-trending folds (oblique folds), of which the Sudbury nappe is one, are distinct from the late folds (longitudinal folds) of the Middlebury synclinorium (Crosby, 1963, p. 116–121; 1964; *see also* footnotes 21, 31, p. 46, 79)—folds of a regime with which thrust faults are most commonly related. Proper interpretation of the Sudbury thrust as a feature of the late regime depends upon its relation to the east limb of the Middlebury synclinorium. The map pattern (Doll and others, 1961) shows that it sharply transects the upturned, westward-topping strata of the east limb, thus indicating a thrust of the late regime. Zen (*in* Cady and Zen, 1960, Fig. 5; Zen, 1961, Pl. 1) extends the Sudbury thrust southward beneath surficial deposits to the Taconic slide, and concludes (1967, p. 34) that the "Sudbury slice . . . is not rooted [in the east limb of the Middlebury synclinorium] but tucks under the main mass of the Taconic rocks," and he suggests "it may well prove to be a parautochthonous sliver, either completely or partly detached, essentially like the sliver at Dorset Mountain" (*see* footnote 33, p. 80). In view, however, of both the lack of exposure of the Sudbury thrust in the vicinity of the Taconic slide and the transection of the Taconic slide by the Dorset Mountain thrust, such an interpretation seems unjustified. Instead, the Sudbury thrust appears to separate Zen's (1967, p. 18–19) "Giddings Brook" and "Sunset Lake" slices, probably once mutually continuous parts of the Taconic allochthon, in much the same way as the Dorset Mountain thrust probably separates his "Dorset Mountain" and "Bird Mountain" slices. Moreover, in view of these relationships and those of thrust and reverse faults that cut the Taconic slide in the Hoosick Falls area, New York (Potter, 1963, p. 61, 64), it seems unlikely that any of the large slices of carbonate rock, possibly including that interpreted (Zen, 1967, p. 31–32) in the Bald Mountain area, Greenwich, New York, resulted from dragging of the soles of westward-moving slices of the allochthon.

[35] The axis of the Green Mountain-Sutton Mountain anticlinorium coincides in Quebec with the axis of the "Quebec Geanticline" of Poole (1967, p. 25–26), a feature that Poole suggests first appeared at the close of the early Paleozoic, later than the appearance of the Vermont-Quebec geanticline.

[36] The major folds here included in the Green Mountain-Sutton Mountain anticlinorium are as presented by Cady (1960, p. 535, Pls. 1, 2, and 3). In some other discussions of the Green or Sutton Mountain "anticline" or "anticlinorium," the axial anticlines are all that have been implied. *See* Christman (1959, p. 44); Christman and Secor (1961, p. 47); Clark and Eakins (1968, p. 171); and Dennis (1960, p. 560, 566).

Monoclinal Flexures

Minor monclinal flexures and related kink layers that dip northwest, and in which the axial surfaces of the late as well as the early minor folds are deformed, are sparsely distributed throughout the Green Mountains, Sutton Mountains, Worcester Mountains, Northfield Mountains and nearby belts in north-central and northwestern Vermont and adjacent Quebec. (*See also* Clark and Eakins, 1968, p. 172; White and Jahns, 1950, p. 205–206.) They are not reported elsewhere, but it seems possible that they will eventually be more widely recognized. The monoclinal flexures are formed in steeply west-dipping or vertical beds and foliation—the limbs of the flexures approach parallel with nonfoliate planes of slip (minor thrust faults) on which rocks to the northwest were displaced upward and to the southeast. The same movement sense is shown by the kink layers, made up of single chevron folds in east-dipping beds and foliation. This movement appears to be general and disregards the position and orientation of previously existing structural features. Its interpretation must await studies specifically directed to the monoclinal flexures and related kink layers, especially to their regional distribution.

Joints

INTRODUCTION

Systematic sets of nearly parallel joints, and also randomly oriented nonsystematic joints, have been ignored or only briefly discussed although they have been recognized by most field workers. The present author has likewise doubted the usefulness of joints in structural interpretation. However, looked at more searchingly, joints seem to yield some significant information.

They are here described as "systematic" (= "regional" or "master") and "nonsystematic" (= "minor"), because some other more specific descriptive genetic designations, such as "strike" or "dip" and "*bc*" or "*ac*" joints, may not apply. The genetic terms, "shear joint" and "tension joint" commonly imply more than can be proved. (*See* Hodgson, 1961a, p. 12–13; 1961b; Price, 1959; Roberts, 1961; Secor, 1965, p. 634–635.)

SYSTEMATIC JOINTS

The most extensive joints form systematically oriented, undeformed, planar sets which dip nearly vertically and most of which cross the trend of the late longitudinal folds and related thrust faults at large angles. Thus they fan out northwestward in the New England salient, striking west-northwest along the axis of the salient but striking to the west or southwest south of the axis and to the northwest and north, north of the axis.

The systematic joints control the pattern of streams, which show the distribution of the joints fairly effectively. The northwest-flowing Winooski and Lamoille Rivers in Vermont, the Missisquoi River in both Vermont and Quebec, and the St. Francis and Chaudière Rivers in Quebec, whose headwaters are mostly in the belt of the Connecticut Valley-Gaspé synclinorium, rise southeast of and cross the axis of the Green Mountain-Sutton Mountain anticlinorium (Pl. 1). The White, Ottauquechee, and Black Rivers in southern Vermont, likewise paralleling joints that transect the late longitudinal folds, flow east from Precambrian basement rocks at the axis of the Green Mountain-Sutton Mountain anticlinorium. Systematic joints rarely have been reported southeast of the axial region of the Connecticut Valley-Gaspé synclinorium. In that area, the drainage shows the patterns of faults and of stratigraphic units in domes and arches and in longitudinal folds.

The systematic joints show in weathered and eroded bedrock exposures as steep northern and southern faces of outcrops, as fissures that transect the strikes of bedding and foliation, and as cliffs and faces of slump blocks in which the dips of bedding and foliation and cross sections of minor folds of the late longitudinal regime are revealed. Fresh exposures, in which slump can be excluded, show little evidence of differential movement of the joint surfaces, either tangential or by separation. The joints are spaced a few inches to a few yards apart, the spacing being the widest in massive rocks. Some studies have suggested that conjugate sets of shear joints and genetically related tension joints are represented. (*See* Christman, 1959, p. 58–59; Christman and Secor, 1961, p. 59–66; Clark, 1964b, p. 46, Fig. 8; Crosby, 1963, p. 32–33, 75; Konig, 1961, p. 76–77; Konig and Dennis, 1964, p. 47–49; *also* Freedman and others, 1964, p. 631–632; Lapham and McKague, 1964, p. 643–645; Secor, 1965, p. 635.)

The systematic joints clearly were formed later than the foliation, porphyroblasts, late longitudinal folds, and monoclinal flexures which they transect. They are quite as evidently succeeded by discordant calc-alkalic Devonian felsic-plutonic and dike rocks that are emplaced partly against the joint surfaces in the bedded rocks (Bethel, Woodbury, and Barre Granites—Balk, 1927, p. 45, 53, 54, 72, 77, 79, "granite of the Black Hills"—Dennis, 1956, p. 50). They are likewise succeeded by unmetamorphosed middle Mesozoic mafic dike rocks, chiefly lamprophyres, that are emplaced mainly in the systematic joints in the bedded rocks, but also in "cross" and "longitudinal" joints in the above-mentioned granitic rocks (Balk, 1927, p. 47–64; Murthy, 1957, p. 97).

NONSYSTEMATIC JOINTS

The nonsystematic joints are less extensive and less conspicuous than the systematic joints. They are variously oriented, in some places curved or irregular rather than planar, they rarely form parallel or concentric sets, and they dip at varied angles. They commonly end at systematic joints or bedding surfaces and bedding joints that they join at relatively large angles (nearly 90°). They probably include many virtual strike joints that dip vertically, are nearly at right angles to the northwest-trending sets of systematic joints, and rudely parallel the late longitudinal folds. They also include other types of joints, many of them strike joints, that approach 90° to bedding in variously oriented folds and therefore strike and dip at varied angles. Strike joints marked in a few places by long north-northeast-trending vertical cliffs might possibly be considered to belong to a systematic set (Christman, 1959, p. 58–59).

The nonsystematic joints are probably late inasmuch as they commonly end at the systematic joints. Like the systematic joints, but less commonly, they locally contain felsic and mafic intrusive rocks.

INTERPRETATION OF THE JOINTS

Interpretation of the joints is complicated by both local and general problems which are only partly solved.

Did the stresses responsible begin before deformation (Hodgson, 1961a); early in deformation specifically as precursors of folding (Parker, 1942); during deformation, more specifically as a part of folding (Christman and Secor, 1961, p. 59–66); or after the climax of deformation reached during folding (Crosby, 1963, p. 125; Konig and Dennis, 1964, p. 47)? The present study is least concerned with joints produced by stresses begun before or early in deformation, inasmuch as most of the joints discussed in this study intersect well-folded strata and are themselves undeformed. Probably some predeformational and early deformational joints are preserved in and west of the belt of foreland folds at the southeast margin of the craton in the Hudson, Champlain, and St. Lawrence Valleys.

The stress relations are in turn complicated by the general problem of distinguishing shear or compressional joints from tension joints. This is a distinction that has thus far only been inferred from the relations of apparently conjugate joint sets in the field, as compared with those explained by structural theory and by models (Becker, 1895; Bucher, 1920–1921), or from actual and assumed relations to folds (Hills, 1964, p. 155–158; Willis and Willis, 1934, p. 99–105), but not from the local relations of individual joints.

In the region of the present study, some conjugate joint sets, the bisectrices of whose acute angles trend at right angles to the axes of longitudinal folds, are possibly shear joints produced by the same compressional stresses that are responsible for the folds. Those conjugate sets with acute bisectrices that are not at right angles to the longitudinal fold axes, however, may be related to a stress field developed later than longitudinal folding. The intersecting sets of joints could actually be intersecting shear and tension joints both related and unrelated to previous folding. This seems particularly likely in the vicinity of the New England salient where, first of all, the regional southeast-northwest stresses responsible for the folds were probably vectored at angles less than a right angle to the confining cratonal margins of the salient. So vectored, these stresses would produce only the half of a set of conjugate shear joints that approached most nearly parallel to the salient margin. Also, northwestward bending of the folds into the salient probably produced tensional stresses across the northwest-trending axis of the salient parallel to the axes of many of the folds. Dominance of joints produced simply by tension on the rocks seems doubtful, however, in view of the greater thickness and confinement of the rocks at depth, especially in the eugeosynclinal zone. These factors suggest that compressional stress, rather than tension, was the cause of systematic jointing.

One of the principal stumbling blocks in the interpretation of joints as shear features has been the failure of the joints to show gross evidence of differential displacement parallel to their surfaces. This failure is possibly explained by sudden release of a small amount of elastic strain energy—the outward pressure of water in rock pores and in incipient and growing joints momentarily exceeds rock pressure (*see* Roberts, 1961; Secor, 1965)—in rocks deformed under both hydrostatic load and the differential compressional stress necessary for shearing. This amount of energy would necessarily have been greatly exceeded under the nonhydrostatic as well as dry conditions of shear or of both shear and tension that are conventionally inferred for most faults and for seemingly gaping joints. It is inferred that under hydrostatic and wet conditions, residual shear stress is stored as fluid pore pressure up to the point that it exceeds the lithostatic pressure by an increment able to open the shear joint an imperceptibly small amount and no more, thus immediately dissipating the accumulated shear stress in its vicinity.

Such residual shear stress is believed chiefly responsible for the jointing in the region of the present study. The hydrostatic conditions were provided by the confining pressure of extremely thick stratigraphic sections, most notably the section of the Cambrian and Ordovician in the New England salient in north-central Vermont, which total at least 50,000 feet. Pore fluids came from rocks beneath that were still undergoing compaction and metamorphism. Release of the stress took place when the thick sections were eroded, and as a result the load was removed. These thick sections and many others much thinner were probably similarly affected. (*See* Price, 1959.) Joints that cut some of the thinner sections were probably caused by conventional initial stresses. Flexural folds and thrust faults provide evidence of shear stress under nonhydrostatic conditions in the foreland belt. Nevertheless, high fluid pressure may be significant in some foreland sections that are deformed only by jointing (Secor, 1965, p. 644).

The nonsystematic joints probably were caused by local less-homogeneous stresses; they appear to be secondary adjustments controlled by the systematic joints.

Discordant Calc-Alkalic Plutons

Discordant and commonly nonfoliate (mainly post kinematic) middle Paleozoic granitic plutons, some of batholithic dimensions, are randomly emplaced in lower and middle Paleozoic bedded rocks that are deformed in both the early and late folds (Pl. 1). They transect bedding, axial surfaces of folds, foliation, and also concordant calc-alkalic plutons (Williams and Billings, 1938, p. 1024, Pl. 3), all of which mark previous quasi-cratonic consolidation of the orthogeosyncline, but they are transected by discordant alkalic plutons; they are therefore subsequent magmatic features (Stille, 1940a, p. 16–20; 1940b, p. 17–20; 1950, p. 10–22). They are calc-alkalic (Billings and Keevil, 1946, p. 810–815), and intermediate to felsic like most of the concordant plutons. They are concentrated near the axis of the New England salient where it crosses the eugeosynclinal zone (Billings, 1955; Cooke, 1951; Doll and others, 1961). They are are probably unrelated to the concordant calc-alkalic plutons.[37] None of the discordant plutons are truncated by unconformities. Contact metamorphic aureoles, with zones of sillimanite, staurolite and andalusite, and garnet, are characteristic (Albee, 1968, p. 330; Billings, 1955; Doll and others, 1961; and Appendix C). These prograde metamorphic aureoles are absent, however, where discordant plutons (post-tectonic units of the New Hampshire Plutonic Series of Page—footnote 37) are emplaced in rocks already regionally metamorphosed to sillimanite grade, most notably the lower and middle Paleozoic bedded rocks of the eugeosynclinal zone southeast of the Somerset geanticline. Some of the latter rocks instead show retrograde metamorphic aureoles (Page, 1968, p. 381) at contacts with the discordant plutons.

The relations to metamorphic rocks discussed in the preceding paragraph show that the discordant plutons were emplaced at varied pressure, high temperature, and high chemical activity, especially of water. (*See also* Woodland, 1963, p. 364.) The overall intensity of these metamorphic conditions was greater than that which prevailed in much of the already metamorphosed and partly cooled bedded rock that the plutons encountered. It therefore seems clear that the magma formed at fair depth where it was perhaps palingenetic. The concentration of the plutons in the New England salient suggests tectonic control of their origin.

The retrograde metamorphic aureoles adjoining the discordant plutons emplaced in rocks already of sillimanite grade are probably unique features of these high-grade country rocks rather than effects of the invading granitic magma. Thus water brought by the magma helped downgrade the dry rocks of the contact zones, something it would not have done to wetter, lower grade country rock (*compare with* Page, 1968, p. 381). As well as geodesiccators, the high-grade country rocks were perhaps also georeducers—relatively low in oxygen—that account for the gray, as opposed to pink, granites of the plutons emplaced in high-grade terranes. Moreover, the absence of sulfide mineral deposits in association with the plutons in the high-grade terranes may reflect the lack of openings in deeply buried rocks—to conduct mineral-bearing solutions and receive the minerals.

FOOTNOTES

[37] The discordant and nonfoliate calc-alkalic granitic plutons and related hypabyssal rocks have been referred along with the concordant calc-alkalic plutons to the New Hampshire Plutonic Series (Billings, 1956, p. 61–64; Doll and others, 1961). It seems reasonable, however, to exclude the discordant plutons from the New Hampshire Plutonic Series and to apply another term because of contrasts in distribution and structural relations discussed here. Especially significant reasons are the relatively few discordant calc-alkalic plutons and their limited area in New Hampshire. The term, "New England Plutonic Series" might be appropriate. (*See also* footnote 25, p. 52.) The discordant plutons include the rocks of the "Late Devonian Plutonic Series" of Page (1968, p. 371, 380–382) and of the *post-tectonic* units of the "New Hampshire Plutonic Series" of Page (1968, p. 378).

Epieugeosynclines[38]

Upper Paleozoic (Mississippian and Pennsylvanian) clastic coal-bearing rocks are superimposed on the orthogeosynclinal belt in southeastern New England and the Maritime Provinces of Canada. They lie unconformably upon rocks of the lower and middle Paleozoic eugeosynclinal zone and are hence referred to as epieugeosynclinal (Kay, 1951, p. 56–57). Although they are preserved only southeast of the region of this synthesis, they probably covered wider areas of the eugeosynclinal zone before they were eroded. (*See also* Cady, 1960, p. 565–566.)

The most complete section in southeastern New England is about 12,000 feet thick. It is composed of conglomerates, sandstones, and shales interbedded with coal seams and is preserved in a synclinorium (Narragansett Basin—Emerson, 1917, p. 52–55; Quinn and Moore, 1968, p. 275–276). The original northwest extent of these rocks and those of the Boston Basin is suggested indirectly by a late Paleozoic regional metamorphic overprint in the Merrimack synclinorium and vicinity. This overprint suggests burial of the middle Paleozoic bedded rocks exposed in the synclinorium by upper Paleozoic rocks.

[38] Nearly synonymous with the term epieugeosyncline are less explicit descriptive and genetic expressions as *backdeep*, *Zwischengebirge*, *post-orogenic basin*, and *nuclear basin*.

High-Angle Faults

INTRODUCTION

Two systems of high-angle faults (Pl. 1), one in the foreland belt of the Champlain-St. Lawrence Valley to the northwest, and another that coincides with the belt of the Bronson Hill-Boundary Mountain anticlinorium near the Connecticut River to the southeast, were possibly formed in one or more closely spaced episodes in the early Mesozoic, or possibly but less likely in the late Paleozoic. They are here designated the Champlain-St. Lawrence faults and Connecticut River faults. Hanging walls of most of the faults contain higher stratigraphic units than do the footwalls, and there is little evidence of strike-slip movement; therefore, most appear to be gravity (normal) faults. The principal faults in each system form longitudinal sets subparallel to the late longitudinal folds. Thus their average strikes are north-northeast south of the axis of the New England salient and northeast north of the axis. Transverse faults perpendicular to, or diagonal to, the strikes of the longitudinal faults are chiefly in the foreland belt.

CHAMPLAIN-ST. LAWRENCE FAULTS[39]

A set of longitudinal faults, most of which are downthrown to the southeast, bound fault blocks that contain Precambrian basement rocks and lower Paleozoic bedded rocks that tilt downward chiefly to the northwest in the valleys of Lake Champlain and the St. Lawrence River. Transverse sets, about half downthrown to the north and the other half to the south, transect the longitudinal high-angle faults and the late longitudinal folds and related foreland thrust faults on the west limb of the Middlebury-Hinesburg-St. Albans synclinorium. (*See* Emmons, 1888, p. 94–98; Fisher and others, 1961; Clark, 1947, Map 642; 1954a, b; 1964a, p. 22; 1964b, p. 52–53; 1966, p. 43–45; Doll and others, 1961; Kumarapeli and Saull, 1966; Osborne, 1956, p. 166; Wiesnet and Clark, 1966). The longitudinal high-angle faults rarely intersect the foreland thrusts and the transverse faults die out eastward, usually within a mile east of the thrust traces, although there are a few exceptions.

The displacement on the Champlain-St. Lawrence high-angle faults decreases both to the northeast and to the southwest, as well as to the southeast of protuberances of the Precambrian basement, most notably the Adirondack

Mountains. Thus the faults are apparently related to development of the domal surface of the Precambrian basement of the Adirondacks (Cady, 1945, p. 571, 579). The exposed basement and the faults extend farthest east of the Adirondacks to an area west of the culmination of the late longitudinal folds between the Middlebury and Hinesburg synclinoria. It is therefore inferred that this culmination, and possibly that between the Hinesburg and St. Albans synclinoria, were produced by the differential vertical movements shown by the gravity faults.

The Champlain-St. Lawrence high-angle faults clearly succeed the late longitudinal folds and the thrust faults that they transect. They also transect middle Mesozoic mafic dike rocks (Hudson and Cushing, 1931, p. 101) and both cut and are cut by felsic dike rocks (Buddington and Whitcomb, 1941, p. 86) probably consanguineous with the mafic rocks. An upper Mesozoic mafic stock (Monteregian pluton) apparently transects one of the longitudinal high-angle faults (Clark, 1954b).

CONNECTICUT RIVER FAULTS

The Connecticut River faults, which are mainly east of the Connecticut River, are longitudinal high-angle faults downthrown principally to the northwest, and that cut folded and metamorphosed lower and middle Paleozoic bedded rocks and middle Paleozoic calc-alkalic plutons. They are chiefly south of the axis of the New England salient, and most of them outline the west side of the belt of middle Paleozoic domal anticlines near and southeast of the axis of the Bronson Hill-Boundary Mountain anticlinorium. These faults (specifically the Triassic border fault—Thompson and others, 1968, p. 215, Pl. 15–1a, 15–1b) extend south of the region of the present study into Massachusetts and Connecticut where they form the east boundary of a taphrogeosyncline that contains eastward-tilted lower Mesozoic (Upper Triassic) bedded rocks. Transverse faults are rare in the northern part of the belt of the Connecticut River faults, but diagonal, northeast trending transverse faults that transect those of the longitudinal set and show apparent transcurrent movement, predominantly right lateral, are common in Connecticut. (*See* Balk, 1956a, 1956b; Billings, 1955; 1956, p. 118–119; Fritts, 1963; Kay, 1951, p. 60; Rodgers and others, 1956; 1959, p. 15–18). The northernmost of the Connecticut River high-angle faults, which extend northeast across the axial region of the New England salient, have been described as thrust faults (Ammonoosuc thrust—Billings, 1956, p. 115–116; and possibly the Northey Hill thrust near Northey Hill—Billings, 1937, p. 530–531). In several places, overturning of units in the hanging walls of one of the faults, apparently by upward drag, seems to indicate that the fault is upthrown to the northwest and must be interpreted as a thrust (Billings, 1933, p. 148–151; 1937, p. 529–530; Lyons, 1955, p. 133). However, the lower Paleozoic rocks southeast of these northern faults are adjoined to the northwest by middle Paleozoic units in the hanging walls that thus appear in plan to be downthrown (Billings, 1955).

It is especially apparent south of the northwestern New England region in Massachusetts and Connecticut, that the Connecticut River faults outline the west side of domal anticlines. In that area, the border fault on the east side of the taphrogeosyncline is in the west flanks of long domal anticlines in the southern extension of the Bronson Hill-Boundary Mountain anticlinorium (Lundgren, 1962, p. 5, Fig. 1). A few high-angle faults, downthrown to the southeast (most notably the Grantham fault—Thompson and others, 1968, p. 215, Pl. 15–1a, 15–1b), are on the southeast flanks of the domal anticlines in New Hampshire (Billings, 1955; 1956, p. 120). It should also be noted that in Massachusetts and Connecticut, gravity faults downthrown to the east that are west of the taphrogeosynclines, are on or near the east flanks of another set of domal anticlines (Fritts, 1962, p. D33-D34; Stanley, 1964, p. 7, Fig. 2); moreover, that a fault downthrown to the west is on the west flank of one of the latter anticlines (Stanley, 1964, p. 86–87).

The Connecticut River high-angle faults clearly succeed both the middle Paleozoic late longitudinal folds (including the domal anticlines) and the concordant and discordant calc-alkalic plutons (Billings, 1955). They are nearly synchronous with the lower Mesozoic (Upper Triassic) taphrogeosynclinal bedded rocks that they bound, especially on the east in Massachusetts and Connecticut. In New Hampshire they precede a lower Mesozoic alkalic stock (Chapman, 1942, p. 1539, Pl. 1) and in Connecticut they precede lower Mesozoic (Upper Triassic?) mafic dikes which nearly parallel and which are apparently intruded along the transverse faults (Rodgers and others, 1956; 1959, p. 15, 17, 18).

OTHER HIGH-ANGLE FAULTS[40]

A few small northeast-trending high-angle faults on the east limb of the Green Mountain-Sutton Mountain anticlinorium in central Vermont show apparent right lateral transcurrent movement (R. H. Jahns and W. S. White, written commun., 1960, *cited in* Doll and others, 1961). They are about half-way between the Champlain-St. Lawrence and Connecticut River faults, and their relationships to these faults, if any, are uncertain.

INTERPRETATION OF THE HIGH-ANGLE FAULTS

The high-angle faults have previously been interpreted in four separate regions with little attempt to interrelate them. The regions are those around Lake Champlain and the city of Quebec to the northwest, and the lower Connecticut River and upper Connecticut River to the southeast. The faults in the Lake Champlain and the city of Quebec regions are probably closely related. Most of them are downthrown to the southeast, and their trend from north-northeast in the Lake Champlain region to northeast along the lower St. Lawrence River near the city of Quebec probably follows the flanks of the New England salient. The interrelations of the faults in the lower and upper Connecticut River regions have been less clear, possibly not only because these regions were separately investigated, but also because the northernmost faults in the upper Connecticut River region had been interpreted as thrust faults, whereas those to the south had been clearly determined to be gravity faults.

The time relations of the high-angle faults require more interpretation than their space relations, particularly because undetermined middle and upper

Paleozoic and Mesozoic bedded rocks, possibly faulted, are eroded from all but the lower Connecticut Valley region where the taphrogeosynclinal lower Mesozoic (Upper Triassic) rocks are preserved. To recapitulate, the faults in the Lake Champlain region both cut and are cut by middle Mesozoic dike rocks. Discordant upper Mesozoic plutons apparently cut the Champlain-St. Lawrence high-angle faults, and discordant lower Mesozoic plutons cut the northernmost of the Connecticut River faults. It is likely that neither faulting nor intrusion was confined to single episodes in the Mesozoic, but it is probable that in each of the regions of high-angle faulting, the principal faults preceded the dominant Mesozoic intrusive activity. Some of the Champlain-St. Lawrence faults show evidence of Precambrian movement (Philpotts and Miller, 1963), and others, perhaps of this fault system, show Middle Ordovician movement (Fisher and Hanson, 1951, p. 808–809).

The spatial relations of the high-angle faults to rocks of low density, especially the granitic rocks exposed in the domal anticlines at the east and west borders of the belt of the Connecticut River faults, suggest the origin of the faults. Downthrow in that area is mainly toward a graben between the two belts of domal anticlines; thus it appears that these belts stood still after late longitudinal folding while the intervening belt subsided. (*See also* Eaton and Rosenfeld, 1960, p. 176–178.) The eastern border fault, at least, was active during sedimentation (Thompson and others, 1968, p. 215) of the taphrogeosyncline about to be discussed.

FOOTNOTES

[39] The Champlain-St. Lawrence high-angle faults discussed here are not to be confused with the "St. Lawrence and Champlain fault," "Champlain-St. Lawrence," or "Champlain-St. Lawrence-Appalachian fault" of various early authors, which were described by Logan (*in* Logan and others, 1863, p. 234) and by Dresser (1912, p. 13), and which referred most explicitly to several of the thrust faults connected with late longitudinal folding. These came to be known as "Logan's Line" (Clark, 1951; Osborne, 1956, p. 170–173).

[40] The "Monroe" and "Northey Hill" hill thrusts are probably relatively minor features that exist only at their type localities. They were extended beyond their type localities principally to help resolve stratigraphic problems that with time have largely disappeared. This explanation seems especially clear in the case of the Monroe thrust (Billings, 1956, p. 97–98; Cady, 1960, p. 551–553).

Taphrogeosyncline[41]

Lower Mesozoic (Upper Triassic) terrestrial clastic rocks as much as 30,000 feet thick unconformably overlie the eugeosynclinal zone of the orthogeosyncline and the epieugeosynclines in southern New England and the Maritime Provinces of Canada. They have been referred to as taphrogeosynclinal (Kay, 1951, p. 60), with special reference to their occurrence in grabens. They (Newark Group) are preserved mainly to the south of the region of the present synthesis where they lie between the Connecticut River high-angle faults (Pl. 1), but they undoubtedly extended north into northwestern New England before they were eroded, much as do the faults (Emerson, 1917, p. 91–100; *see also* Cady, 1960, p. 566; Rodgers and others, 1959, p. 15–18; Sanders, 1963).

Hypabyssal mafic rocks and minor mafic volcanic rocks are included in the Connecticut Valley taphrogeosyncline (Emerson, 1917, p. 95–96, 261–277). They are probably represented in some of the unmetamorphosed mafic dikes in the northwestern New England region. These igneous rocks are interbedded with and emplaced in relatively undeformed bedded rocks that unconformably overlie the fully consolidated orthogeosyncline, hence they are final magmatic features (Cady, 1960, p. 566; Stille, 1940a, p. 20–23; 1940b, p. 20–21).

[41] Nearly synonymous with the explicit term *taphrogeosyncline* are some of the uses of such terms as *rift valley* or *rift geosyncline*.

Discordant Alkalic Plutons

Two curvilinear belts of discordant and nonfoliate plutons enter the region of the present study, where they transect lower and middle Paleozoic rocks and structural features and the Mesozoic high-angle faults (Pl. 1). One of the belts, which contains lower Mesozoic plutons (Jurassic or Triassic White Mountain Plutonic Series) of batholithic dimensions, extends north-northwestward in the orogenic belt. It enters the region from the southeast in northern New Hampshire and crosses the upper Connecticut River into northeastern Vermont. The other belt, which contains upper Mesozoic plutons (Cretaceous Monteregian plutons), enters the region from the Canadian Shield to the northwest near Montreal. It extends eastward into the orogenic belt a little north of the international boundary, and north of the north end of the belt of the White Mountain plutons. Gravity and magnetic anomalies suggest that a comparable and parallel belt of alkalic plutons is at depth in northeastern New York (Diment, 1964). Two lower Mesozoic plutons, one of them the classic Mount Ascutney (Daly, 1903), form a lesser east-west belt west of the Connecticut River in southern Vermont. (*See* Billings, 1955; Cady, 1960, p. 566; Dresser and Denis, 1944, p. 455–482; Doll and others, 1961; Quebec Department of Mines, 1957.)[42]

These discordant plutons are alkalic and mafic to felsic (Billings and Keevil, 1946, p. 801–810; Dresser and Denis, 1944, p. 451). Mesozoic dikes of similar composition are associated with them and also occur widely in intervening areas.

Both the plutons and dikes are clearly non-geosynclinal and non-orogenic, inasmuch as the belts in which they occur trend across all previously consolidated tectonic features of the region including part of the craton. These belts roughly parallel the axis of the New England salient, which suggests crustal weakness common to both. Their linearity suggests that they are controlled by deep crustal fractures (Chapman, 1963) and that the source of the magmas was well below any rocks that crop out. Moreover, it suggests that the magmas were products of crystallization differentiation of basaltic magma, aided by diffusion and assimilation. (*See also* Chapman and Williams, 1935; Swift, 1966, p. 9–11.)

Contact metamorphic effects are commonly limited to a zone of hornfels or marble only a few hundred feet thick in rocks little touched by regional metamorphism (Dresser and Denis, 1944, p. 466, 468). None of these contact zones have been mapped, which is in contrast with the comparatively wide zones that adjoin the discordant calc-alkalic plutons. The bedded rocks adjoining the alkalic plutons were apparently too dry to be contact metamorphosed beyond the range of magmatic water, inasmuch as they were already metamorphosed, or if not, were chiefly limestone or quartz sandstone.

FOOTNOTES

[42] The rocks of the White Mountain Plutonic Series and Monteregian plutons were all assigned to the upper Paleozoic in an earlier publication (Cady, 1960, p. 566), but recent revisions of the geologic time scale (Kulp, 1961) plus new radiometric dating as well as paleomagnetic studies, have made it necessary to reassign these rocks to the Mesozoic (Fairbairn and others, 1963; Faul and others, 1963, p. 16; Hurley and others, 1961a, p. 156; Larochelle, 1968; Lowdon, 1960, p. 38; Opdyke and Wensink, 1966; Tilton and others, 1957, p. 362, Tables 3, 4, 5). They were also erroneously called subsequent magmatic features, in that they were considered to be intrusive rocks emplaced soon after the synorogenic plutons and likewise to be peculiar to the orthogeosynclinal belt (Cady, 1960, p. 565–566). Their newly assigned younger age, plus the further recognition that the Monteregian plutons are probably descendents of magmas similar to nonorogenic continental or oceanic basalt (Fairbairn and others, 1963), reinforces the conclusion that they are cratonic rather than orthogeosynclinal features.

Chronology

INTRODUCTION

The tectonic evolution of the region of northwestern New England and adjacent Quebec is here summarized chronologically, and supporting biostratigraphic, radiometric, and related structural data are presented. The biostratigraphic data provide the principal evidence for the times of sedimentation, volcanism, igneous intrusion, folding, and faulting; whereas the radiometric data, chiefly the K-Ar ages show the times of regional metamorphism and subsequent erosional unloading, or are hybrids of separate events of this sort.[43] Both the biostratigraphic and radiometric data in general indicate successively more youthful bedded and metamorphic rocks toward the southeast, though the metamorphic ages are less than the biostratigraphic ages. Thus, the zone of southeastward transition from the belt of early Paleozoic metamorphism to that of middle Paleozoic metamorphism near the axis of the Middlebury-Hinesburg-St. Albans synclinoriums in Vermont and Quebec, is well within the terrane of biostratigraphically determined lower Paleozoic rocks. Likewise the zone of southeastward transition from the belt of middle Paleozoic metamorphism to that of late Paleozoic metamorphic ages, a little northwest of the Bronson Hill anticline in New Hampshire, is within the terrane of middle (and also lower) Paleozoic rocks.

Most biostratigraphic data (some still controversial) come from the early and middle Paleozoic miogeosynclinal zones. The environment of these zones was most favorable for the organisms, and the fossils were neither diluted by inorganic clastic constituents nor destroyed by metamorphic differentiation. The graptolites preserved in the least metamorphosed rocks of the eugeosynclinal zone are very helpful. Stratigraphic sequences are most readily recognized in the miogeosynclinal zone where persistent, thin, and virtually monomineralic units least subject to metamorphic differentiation are common. Correlation is helped locally where fossiliferous rocks of miogeosynclinal type persist laterally into the eugeosynclinal zone. Lateral facies changes obscure correlations between the miogeosynclinal and eugeosynclinal zones and the age relationships are uncertain in the latter because many transitions are into thick monotonous units of structurally and metamorphically complex rock. Unconformities of regional or interregional extent support comparisons of chronologies in separate structural belts. Thus, widespread uplifts that include the several geanticlines and the neighboring foreland belt are inferred. (*See* Cady, 1968b.)

The radiometric data are still being evaluated, and much of it reflects chiefly the times of cooling of metamorphic and intrusive rocks, marked stratigraphically by the several unconformities in the lower and middle Paleozoic. The unconformities show that erosional unloading and cooling were several successive events.[44] Some of the metamorphic and intrusive rocks, especially in the eugeosynclinal zone, doubtless were formed earlier than the radiometric data indicate because of their greater depth of burial and the correspondingly longer time before unloading and cooling. The contact relations of radiometrically dated intrusive rocks emplaced in paleontologically dated bedded rocks or their correlatives leave much room for inference of dates of intervening sedimentary and igneous events.

The metamorphic age of a Paleozoic bedded rock may range from a few tens of million years to as much as 200 m.y. younger than its biostratigraphic age, depending respectively upon whether the radiometric systems record, (1) episodes of simple uplift, erosional unloading, and the close of static metamorphism in thick sections in the geosynclinal troughs; or (2) much later erosional unloading, following orogeny and possibly connected dynamothermal metamorphism or igneous intrusion, or both. Radiometric ages older than the biostratigraphic ages of the various Paleozoic rock units have not been reported, therefore it seems unlikely that old detritus or xenoliths have significantly contaminated the samples. Hybrid metamorphic ages are common, thus Late Ordovician ages that seem to indicate folding and metamorphism of the Green Mountain-Sutton Mountain anticlinorium in the Taconic disturbance instead of the Acadian orogeny, are in fact a few among many apparent ages evenly spread from Late Cambrian to Devonian Acadian and even later times.

The regional foliation conveniently separates in time the similar folds of the early (pre- and early Acadian) regime, whose axial surfaces it parallels, from the domes and arches and folds more nearly of parallel style, by which it is in turn deformed in the late (Acadian) regime. This foliation formed in several episodes before the end of the early fold regime. Thus the youngest foliation in the northwestern part of the eugeosynclinal zone, which is truncated by an unconformity that marks stabilization at the end of the early Paleozoic, is older than foliation to the southeast where the eugeosynclinal zone and its potentially foliate rocks persisted almost until consolidated in the late regime in the middle Paleozoic.

The following discussion is broken down into the biostratigraphic intervals between major unconformities, inasmuch as the biostratigraphic record more faithfully reflects the tectonic history than does the radiometric record. Unconformities beneath the lowest bedded rock units (Precambrian) and above some of the higher units (middle Paleozoic) are known only outside the region.

PRECAMBRIAN

The Precambrian chronology in the region is inherently scanty. It is pieced together from a few radiometric values obtained from the basement rocks exposed in the Green Mountain-Sutton Mountain anticlinorium in Vermont

(Mount Holly Complex) and by lithologic correlation with the neighboring basement in the Precambrian of the Adirondack Mountains and of the Grenville province of the Canadian Shield (Fig. 1). The basement in the Grenville province has been variously dated as early, middle and late Precambrian[45] depending upon interpretations of original sedimentary facies relationships, intrusive relationships, and metamorphic overprints (Appendix E). Eugeosynclinal strata in the northwestern part of the Grenville province were probably deposited in the early Precambrian, but the widespread miogeosynclinal rocks in the southeastern part, in the Adirondack Mountains, and in the Green Mountains, are of middle or late Precambrian age, or both. The dating is obscure because the most commonly used radiometric systems (K-Ar) were reset about 1 b. y. ago (Osborne and Morin, 1962, p. 135, 201; Stockwell, 1961, p. 113).

Radiometric ages of zircon (Pb/alpha) reported from the basement rocks of the Green Mountain-Sutton Mountain anticlinorium and adjacent domes to the east in southern Vermont, range from 900 to 1530 m.y. (Faul and others, 1963, p. 3, 7). Three of the zircon ages (900, 970, and 1100 m.y.)—one (970 m.y.) clearly from bedded basement rock—and a muscovite age (Rb-Sr) (928 m.y.) from pegmatite (Faul and others, 1963, p. 7), probably record the event that has been called the "Grenville orogeny" in the Adirondacks and elsewhere, and fix a maximum age of very late Precambrian sedimentation during which detrital zircon was deposited. This event is here referred to as a 1.1-b.y. cooling event or simply and more loosely as the billion-year event.[46]

The K-Ar values reported from the Precambrian basement rocks, ranging from 859 to 318 (Table 1) are hybrids of the billion-year event and at least two episodes of Paleozoic metamorphism and subsequent cooling.

Stratigraphic correlation of the bedded Precambrian basement rocks in the Green Mountain-Sutton Mountain anticlinorium with those exposed in the Adirondack Mountains is suggested by lithic similarity (Brace, 1953, p. 28; Engel, 1956, p. 89–90, 93; M. S. Walton, written commun., 1963). A zircon (U-Pb) isotope study that apparently is able to penetrate the billion-year overprint in bedded basement rocks in the Adirondacks has yielded an age of 1220 ± 15 m.y. (Silver, 1965). This late Precambrian age is between the two dates (Pb/alpha), 970 and 1530 m.y., cited above, from the late Precambrian basement of the Paleozoic orthogeosynclinal belt in Vermont.

The Precambrian record of the northwestern New England-Quebec region closes with little more than differential uplift, and connected erosional unloading and cooling that restarted the potassium-argon systems, about a billion years ago. (*See also* MacIntyre and others, 1967, p. 826–827.)

EARLY PALEOZOIC

Chronologic Relationship to the Precambrian

The early Paleozoic chronology starts with an uncertainty about both the age and the stratigraphic correlation of the lowermost bedded rocks. These rocks, unconformable on the Precambrian basement complex, lie conformably beneath the lowest fossiliferous Lower Cambrian in the orthogeosyncline (*see also*

Poole, 1967, p. 17–18). They are assigned the age of Cambrian(?), which is the usage of the U.S. Geological Survey.[47] Their true age is masked by the overprints of one or more episodes of subsequent metamorphism in the Paleozoic (Pl. 1). Transecting intrusive igneous rocks (mainly lamprophyre dikes) are of Mesozoic age, so offer no assistance. The oldest metamorphic age, to be discussed in a later section, is 530 ± 35 m.y., as compared with the billion-year event of the late Precambrian.

Biostratigraphic Chronology of the Early Paleozoic

The chronology of the biostratigraphically established lower Paleozoic—Cambrian and Ordovician—bedded rocks includes the following paleontologic record. In the miogeosynclinal zone are Early, Middle, and Late Cambrian and Early and Middle Ordovician fossils (*cited by* Cady, 1960, p. 553–554; *see also* Shaw, 1966b; Theokritoff, 1968, p. 13–15).[48] In the eugeosynclinal zone are Early, Middle, and Late Cambrian and Early, Middle, and Late Ordovician faunas (*see* Berry, 1962; Bird and Rasetti, 1968; Boucot and Drapeau, in press; Boudette and others, 1967, p. 24–27; Cady, 1960, p. 554; Harwood, 1968, Harwood and Berry, 1967, p. D20; Lamarche, 1965, p. 155–157; Osborne, 1956, p. 175, 182, 191, 192; written commun., 1964; Osborne and Berry, 1966; Osborne and Riva, 1966; A. R. Palmer, written commun., 1965; Riva, 1966b; St. Julien, 1963b, p. 3; 1965; Theokritoff, 1964; 1968, p. 15–18; Zen, 1964b, p. 18, 20, 21, 24, 26; *see also* Berry, 1968, p. 23–24, 26, 29, 30, 31). And in the exogeosyncline the fossils are Late Ordovician (Clark, 1955, p. 19–36; 1964a, p. 5–10; 1964b, p. 34–39; 1964c, p. 8–62).

The chronology of the lower Paleozoic eugeosynclinal zone is also tied to the tracing of a distinctive stratigraphic unit, the Ottauquechee Formation, that extends along the southeast limbs of the axial anticlines of the Green Mountain-Sutton Mountain anticlinorium (Doll and others, 1961). This unit (Pl. 2) has been followed northwestward in Quebec across the axial anticlines into the Middle and Upper Cambrian Sweetsburg Formation (Doll and others, 1961) of the miogeosynclinal zone (Cady, 1960, p. 548, 554; 1968b, p. 157; Osberg, 1956). It also may be traced northeast in Quebec in the eugeosynclinal zone along the flanks of the anticlinorium as the L'Islet or Rosaire Formation (Béland, 1957, p. 15; 1962, p. 12–14; Cady, 1960, p. 548–549; 1968b, p. 157; Dresser, 1912, p. 20–22; 1913, p. 51; Dresser and Denis, 1944, p. 389; Knox, 1917, p. 234). It is, in addition, comparable lithically and sequentially to a distinctive Middle[49] and Upper Cambrian unit, the Hatch Hill Formation, in the eugeosynclinal rocks of the Taconic klippen (Cady, 1968b, p. 157; Doll and others, 1963, p. 95; Theokritoff, 1964, p. 178–179; 1968, p. 16; Zen, 1964a, p. 42–43; 1964b, p. 21–22; 1967, p. 53–54).

The Ordovician of the eugeosynclinal zone also contains a distinctive stratigraphic unit, the Moretown Member of the Missisquoi Formation (Doll and others, 1961). This unit may be traced northeastward along the east limb of the Green Mountain-Sutton Mountain anticlinorium, across a facies boundary (near the international boundary), and into a section of at least two stratigraphic units in Quebec that near its top includes fossiliferous upper Middle Ordovician (Berry,

1962).[50] It is nearly identical to rocks of the upper part of the Ascot Formation (Lamarche, 1967, p. 3–5; St. Julien, 1965; St. Julien and Lamarche, 1965, p. 6–8) that crop out farther northeast in the Stoke Mountain anticline. The Moretown Member and the overlying Cram Hill Member of the Missisquoi Formation are also useful in extending the chronology into southeastern areas of the eugeosynclinal zone, southeast of the Connecticut Valley-Gaspé synclinorium (Cady, 1968b, p. 158–159). In the southeastern areas, sequences of units lithically almost identical to the members of the Missisquoi Formation, most notably the Albee, Ammonoosuc, and Partridge Formations (Billings, 1955; Doll and others, 1961) and the Albee and Dixville Formations (Green, 1964, p. 23–32; 1968, p. 1607–1613; Green and Guidotti, 1968, p. 257–259; Harwood and Berry, 1967, p. D17–D20; Hatch, 1963, p. 12–18), are widely known (Cady, 1960, p. 550–551). The Dixville Formation includes fossiliferous upper Middle Ordovician (Harwood and Berry, 1967, p. D20) near its top. The Ordovician chronology of the eugeosynclinal zone, like that of the Late Cambrian, is further supported by the presence of lithically and sequentially comparable fossiliferous Middle Ordovician west of the Green Mountain-Sutton Mountain anticlinorium (Cady, 1968b, p. 157). Lithically comparable sequences of rocks are found west of the anticlinorium in the miogeosynclinal Stanbridge Slate (Cady, 1960, p. 550; Osberg, 1956; Riva, 1966a) and in the eugeosynclinal Mount Hamilton and Pawlet Formations (Doll and others, 1961) of the Taconic klippen (Zen, 1961, p. 331–333; 1964a, p. 51–52, 62–64).

The biostratigraphic record of the early Paleozoic is broken internally by unconformities (already discussed), recognized chiefly within and beneath Middle Ordovician rocks. Except locally, the record is apparently also broken near the top of the Ordovician in the eugeosynclinal zone. Fossiliferous bedded rocks of Late Ordovician age (Sherbrooke Formation) have been found only northwest of the Stoke Mountain geanticline and vicinity (Pl. 2) (Lamarche, 1965, p. 155–157; St. Julien, 1965). They were once probably more extensive, but were eroded at the time of the Taconic disturbance. (*See also* Boucot, Field and others, 1964, p. 88–90).

Radiometric Chronology of the Lower Paleozoic Rocks

Radiometric dates in the lower Paleozoic rocks and in closely subjacent Precambrian basement rocks, indicate respectively prograde and retrograde metamorphism in the orthogeosynclinal belt, mainly during early to middle Paleozoic times, although late Paleozoic dates are represented in eastern areas (Tables 1 and 2). They mark restarting chiefly of potassium-argon systems in muscovite, biotite, and whole-rocks at a critical point of cooling, ideally only in the latest episode of metamorphism. Fifty-four K-Ar values from biostratigraphically dated metamorphosed lower Paleozoic sedimentary and igneous intrusive rocks at 47 localities (Table 2) range from 481 to 245 million, corresponding to a spread of more than 200 m.y. that includes the interval from perhaps the Early Ordovician well into the Permian. The oldest 14 ages, ranging from 481 m.y. to 414 m.y., correspond to the time interval of the late early Paleozoic and early middle Paleozoic (*see also* Bottino and Fullagar, 1966, p.

1173). They thus appear Ordovician or Silurian, or both, and the younger 11 of these (446 and less) suggest the close of regional metamorphism presumably connected with the Taconic disturbance (Harper, 1968a, p. 56; 1968b; 1968c; Lowdon, 1963a, p. 105; 1963b, p. 80; Rickard, 1965, p. 530–534). Five of these eleven values, 441 ± 15,440 ± 30,439 ± 8, 433 ± 13, and 430 ± 20, variously approximate the interpreted radiometric dates of the close of the Ordovician (440 m.y.— Holmes, 1959, p. 204; 425 m.y.—Kulp, 1961, p. 1111) and also the time of the Taconic disturbance; but more detailed radiometric data based on collections of statistical numbers of samples are necessary to show a possible Taconic peak of metamorphism. It seems probable that the ages, 423–414 m.y. inclusive, are at the most, hybrids (Appendix G) between early Paleozoic (Cambrian and Early Ordovician) regional metamorphic events and middle Paleozoic (Devonian) metamorphic events that reflect partial loss of argon during middle Paleozoic metamorphism. (*See* Church, 1968; Dennis, 1968, p. 959, *also* Long and Kulp, 1962, p. 981–982, Fig. 3, p. 983.) The values, 407, 405, 395, 393, 391, and 390, which correspond approximately to the early Devonian, certainly represent either hybrid or true post-Taconic ages.

A high K-Ar value of 530 ± 35, which probably reflects little loss of argon and therefore may be significant in the foregoing discussion, has been obtained from metamorphic rocks at a locality in the Gaspé Peninsula, Quebec (Fig. 1), northeast of the region of the present study (Lowdon, 1963a, p. 106–107). It does not correspond to the Late Ordovician and Early Silurian interval of the Taconic Disturbance even by applying the full limit of analytical error (*see also* Ollerenshaw, 1968, p. 17–18). Instead, the value 530 corresponds to the boundary between the Middle and Late Cambrian (Kulp, 1961, p. 1111) and is probably not a hybrid metamorphic date. The dated metamorphic rocks (Shickshock Group) are just beneath an apparently angular unconformity with overlying unmetamorphosed Silurian rocks (McGerrigle, 1954, p. 36; Ollerenshaw, 1968, p. 94), an unconformity that may be referred to loosely as "Taconic" (Boucot, and others, 1964, p. 90; Pavlides and others, 1968, p. 61–63). The fact that the Silurian and younger rocks are unmetamorphosed, though folded and faulted, could explain preservation of an early K-Ar date. (*See also* Lajoie and others, 1968, p. 621; Lespérance, 1959, p. 2.)

To the southwest of the Gaspé Peninsula in the northwestern New England-Quebec region where the change southwestward along the strike is from unmetamorphosed to metamorphosed middle Paleozoic rocks (Pl. 1), the K-Ar dates obtained from the lower Paleozoic strata suggest only Ordovician and younger metamorphic ages (Table 2) rather than a Cambrian age such as was found in the Gaspé. Thus it is probable that the Cambrian K-Ar record has been partly to completely destroyed to the southwest because of expulsion of argon during middle Paleozoic and later metamorphism. Additionally significant in this connection is the fact that the 14 K-Ar values discussed above—481 to 414 corresponding to the interval from Early Ordovician perhaps through Silurian—are from northern and also western areas. Lower values are from southern and eastern areas, where the rocks are more highly metamorphosed and where more thorough degassing of the Cambrian micas would be expectable.

The 530 ± 35 m.y. date of the Cambrian micas in the Gaspé suggests either a previously unrecognized orogenic episode, or in the absence of clear evidence of orogeny, merely a critical cooling as a response to uplift and erosion of eugeosynclinal rocks regionally metamorphosed at depth. Some younger early Paleozoic K-Ar dates also probably reflect critical cooling simply in response to uplift and erosion. The early and middle Paleozoic dates thus appear to be separated by at least one episode of critical cooling that took place as stratigraphic units and also recumbent folds and slides accumulated upward in the subsiding geosynclinal troughs. This interpretation seems to be supported by the occurrence of randomly oriented foliate clasts in Middle Ordovician sedimentary breccias (Table 2) (Aniruddha, 1968; Lamarche, R. Y., 1969, written commun.; Riordon, 1957, p. 390–393; Wanless and others, 1966, p. 79); the foliation in the clasts shows a probable hybrid Late Ordovician K-Ar value of 441 ± 15 (Wanless and others, 1966, p. 78).

The Paleozoic metamorphic overprints have been circumvented in several places by Rb-Sr whole-rock studies.

One of the values obtained, the highest radiometric value from the region of the present study, 494 ± 20, is an Rb-Sr whole-rock determination of a Cambrian (?) sericite phyllite interbedded with greenstone (Appendix H). The protolith of the phyllite is rhyolite and that of the greenstone is basalt. These two form a stratigraphic unit, the Tibbit Hill Volcanic Member of the Pinnacle Formation (Doll and others, 1961)—contrasting volcanic rocks are apparently differentiated from the same magmatic source. This value, though larger than the K-Ar values (460 and less) discussed above, still does not reflect complete absence of metamorphic overprint, though it properly should indicate time of magmatic differentiation in the prefossiliferous Early Cambrian (Cambrian?). Instead, it corresponds to Late Cambrian and Early Ordovician on the time scales (*see also* Harris and others, 1965). It is, moreover, too great to match with the K-Ar metamorphic value (Table 2) of 420 ± 30 from muscovite in nearby Cambrian (?) schistose graywacke (Pinnacle Formation) (Lowdon, 1963a, p. 104).

The metamorphic overprint has possibly also been penetrated, though to a lesser extent, by Rb-Sr investigations southwest and south of the region of the present study. In the Hudson Valley in southeastern New York, an Rb-Sr muscovite age combined with an Rb-Sr whole-rock analysis gives an apparent value of 430 ± 20 m.y. (Long, 1962, p. 1004), or perhaps Late Ordovician (Kulp, 1961, p. 1111). This is possibly only 15 m.y., but perhaps as much as 30 m.y., later than deposition of the Middle Ordovician (*compare* Kulp, 1961, *with* Holmes, 1959) unit (Hudson River Pelite) from which it was obtained. A similar Rb-Sr whole-rock age of 440 ± 30 m.y. reported from western New Hampshire, in the region of the present study (Naylor, 1969, p. 415), is about the age of the Middle Ordovician (Kulp, 1961, p. 1111) protolith (Ammonoosuc Volcanics), and if this protolith is indeed 460 ± 15 m.y. old (*compare with* Holmes, 1959) as seems possible from its Rb-Sr dating to the south in Massachusetts (Brookins, 1968), the ages from New Hampshire are perhaps 20 m.y. less.

Other dates that penetrate an extensive late Paleozoic metamorphism and K-Ar overprint are reported in southeastern New England. Gneisses in eastern

Connecticut (Brookins and Hurley, 1965), dated by the Rb-Sr whole-rock isochron method, are as much as 472 m.y. old (or early Middle Ordovician). New Pb-alpha determinations of zircon and monazite, also from gneiss in the same general area, give early Paleozoic ages of 530 ± 60 m.y. to 450 ± 50 m.y. (Snyder, 1964, Table 2, p. 28; *see also* Stern and Rose, 1961). A foliated pegmatite in this area is more than 500 m.y. old (Late Cambrian), as shown by uranium-thorium-lead isotopic analysis of zircon (Zartman, Snyder and others, 1965, p. D5-D6).

Clearly intrusive igneous rocks and their contact metamorphic aureoles, wholly or partly northwest of the middle Paleozoic metamorphic overprint in Quebec (Pl. 1) furnish K-Ar dates useful in interpretation of the early Paleozoic chronology of the orthogeosynclinal belt. The two oldest such dates—both 495 ± 35 m.y.—were obtained from muscovite and biotite in a garnetiferous micaceous quartzite (Shickshock Group), in the contact metamorphic zone of an ultramafic pluton in the Gaspé Peninsula, Quebec. The bedded rocks are evidently of pre-Ordovician age, and the pluton of Late Cambrian or Early Ordovician age (Lowdon, 1963a, p. 107–108). This occurrence suggests that other hot ultramafic intrusions capable of contact metamorphic action were emplaced elsewhere in the orthogeosyncline at about the same time, although none that have been exposed in the region of the present study show clearly thermal metamorphic contact zones that might be dated radiometrically.

Two more K-Ar dates (Table 2)—477 to 481 m.y.—have been obtained from muscovite in dikes of cataclastic and recrystallized pegmatitic granite and quartz monzonite that are emplaced in ultramafites southeast of the Green Mountain-Sutton Mountain anticlinorium in Quebec (Poole and others, 1963, p. 1063). The ultramafic and associated mafic rocks in this area intrude strata of the interval of Upper Cambrian to probably lower Middle Ordovician (Caldwell Formation). Their penetration of this interval of course does not preclude the possibility that they also intrude upper Middle Ordovician or higher rocks in the region, but it simply indicates that some, and most in the areas in which those containing the felsic dikes are observed, intrude Ordovician rocks that might be no younger than about 480 m.y.

Perhaps these ultramafites are as old as those of the Gaspé Peninsula that are discussed above, and the K-Ar values from the felsic dike rocks are hybrids of two episodes of critical cooling about synchronomous with those in the bedded rocks. Accordingly, those two episodes were each followed by cataclasis and recrystallization that accompanied deformation; the first episode would have been well before known Middle Ordovician rocks were deposited, and the second in the Devonian (Acadian). Possibly, however, the K-Ar values are nearly true ages that reflect cooling soon after erosional unloading of the felsic dikes and before unconformable superposition of the Middle Ordovician. Perhaps, also, the ultramafites were emplaced at a later time than in the Gaspe, and after cooling. If so, the felsic dikes possibly intruded them and were critically cooled before the enclosing ultramafic rocks were displaced to their present sites (Appendix I). [51] (*See also* Wanless and others, 1966, p. 79–80.)

The Middle Ordovician or earlier dates of ultramafic intrusion in the northernmost parts of the Appalachian belt (just discussed), are equalled by comparable K-Ar dates of 455 to 465 m.y. in the central Appalachians in Pennsylvania (Lapham and Bassett, 1964, p. 665).

Zircons from a granodiorite pluton emplaced in Ordovician rocks southeast of the Bronson Hill-Boundary Mountain anticlinorium in northern New Hampshire and adjacent Maine, have yielded a Pb-alpha age of 450 ± 50 m.y. (Green, 1964, p. 44–46)—about Middle Ordovician. Zircons from a sample taken from the foliate border zone of the pluton gave a Pb-alpha age of 360 ± 40 m.y., thus reflecting the middle Paleozoic (Devonian) metamorphic overprint. Such ages are subject to ambiguous interpretations. A yet later overprint of late Paleozoic (Mississippian) age is suggested by a sample of the pluton that provided a K-Ar age of 323 m.y.

An Rb-Sr whole-rock date of 440 ± 30 m.y. and a Pb^{207}/Pb^{206} zircon date of 450 ± 25 m.y., Middle (?) Ordovician (*see* Kulp, 1961, p. 1111)—has been obtained from granite (probably of the Highlandcroft Plutonic Series) exposed at the cores of the domal anticlines in the Bronson Hill-Boundary Mountain anticlinorium in western New Hampshire (Naylor, 1968, p. 234; 1969, p. 420).[52] The bedded rocks, which include regional metamorphic granitic rocks exposed in these anticlines, are Middle Ordovician and possibly older. A pluton of comparable biostratigraphic age (Albee, 1961) at the northeast end of the Bronson Hill-Boundary anticlinorium in Maine has provided K-Ar values of 373, 366, and 356 m.y. that reflect the middle Paleozoic metamorphic overprint and one of 433 m.y. (F. C. Canney, oral commun., 1968) that is possibly within a few m.y. of the time of intrusion and related mineralization (Table 2).

Hornblende from granite (Quincy Granite) in easternmost Massachusetts (Zartman and others, in press) shows an early Paleozoic K-Ar age of 442 ± 22 m.y. (Middle or Late Ordovician) amidst a sporadic late and possibly middle Paleozoic overprint that becomes more continuous northwestward toward the axis of the orthogeosynclinal belt. This relationship, which is the symmetrical opposite of that in northwestern New England and adjacent Quebec where the younger overprints appear successively southeastward from the northwestern margin of the orthogeosyncline, suggests that the rocks in eastern Massachusetts that provide early Paleozoic radiometric ages are near the southeastern margin of the orthogeosynclinal belt (*see also* footnote 4, p. 30).

The youngest early Paleozoic date provided by clearly intrusive igneous rocks —420 ± 30 m.y.—was obtained from biotite in a granite pluton (Jacques Cartier Granite) that intrudes Lower Ordovician sedimentary rocks in the Gaspé Peninsula (Lowdon, 1963b, p. 81). This pluton could fall in Late Ordovician or Early Silurian.

Summary of Early Paleozoic Tectonic Events

The orthogeosyncline, including the New England salient, was established in the earliest Paleozoic, or possibly the latest Precambrian. The eugeosynclinal zone at first occupied the full width of the orthogeosyncline, but before the fossiliferous Early Cambrian its western limits in Vermont and immediately adjacent parts of Quebec retreated eastward from the cratonal margin to the Vermont-

Quebec geanticline, leaving the miogeosynclinal zone in and south of the Champlain Valley. In more distant areas to the north in Quebec, however, the northwestern edge of the eugeosynclinal zone remained at the margin of the craton until Middle Ordovician time. (*See* Pl. 3.)

The Vermont-Quebec geanticline, at least started to form during the Cambrian by tectonic stillstand relative to subsidence of adjoining geosynclinal tracts. The other geanticlines to the southeast probably first appeared in the Early to Middle Ordovician. Subsidence was more rapid in the eugeosynclinal zone than in the miogeosynclinal zone. It was most rapid where the eugeosynclinal troughs cross the axis of the New England salient—here the rate of subsidence was two to three times that on the flank of the salient. As the troughs subsided, volcanics were extruded and sediments derived from the craton, geanticlines, volcanic accumulations, and from intra-eugeosynclinal uplifts (Pl. 3) (Appendix A), were deposited in, and to a less extent, over the geanticlines. Turbidity currents, slides, and isoclinal recumbent folds moved down the slopes, especially of the uplifts, in response to gravity. In some places, the cumulative movement (commonly to the northwest) was upslope, and where the folds were blocked, those folds face upward, as on the crest of the Vermont-Quebec geanticline near the international boundary.

Ultramafic, mafic, and intermediate hypabyssal and plutonic rocks were the first to be emplaced, penetrating the bedded rocks now exposed in the eugeosynclinal zone possibly at the end of the Cambrian or beginning of the Ordovician. Contact zones of rodingite developed at the borders of dunite and peridotite. Thereafter, the solid serpentinized dunite and peridotite were transported chiefly northwestward, urged on by the pressure of thickly accumulated overlying geosynclinal deposits and possibly by tectonic overpressure as serpentinization continued and plasticity increased. Albitic granitic plutons were possibly emplaced a little later in the Early or Middle Ordovician near the Stoke Mountain geanticline. Regional foliation, which had first appeared in the Cambrian as the geosynclinal troughs subsided, was formed parallel to the axial surfaces and limbs of the early folds, and was superimposed upon previously formed ultramafic to felsic plutonic rocks, including the granitic rocks just mentioned.

Stillstand of the Vermont-Quebec and Stoke Mountain geanticlines and subsidence of the geosynclinal troughs gave way to episodes of general uplift of the orthogeosyncline in the Late Cambrian to Middle Ordovician. These uplifts led to denudation mainly of the geanticlines as evidenced especially by the Middle Ordovician unconformities and the Taconic slide (Pl. 3). Rocks containing some of the early folds and the regional foliation, and the various plutonic rocks discussed above, were eroded, especially from the Stoke Mountain geanticline. The Taconic allochthon slid westward from the southern part of the Vermont-Quebec geanticline, probably urged on by a "head" of thick water-filled and rising eugeosynclinal deposits to the east of the geanticline.

The tectonic disturbances just discussed that culminated in the Middle Ordovician, might be called the early or Stoke Mountain phase of the Taconic disturbance.[53] This designation seems appropriate, because in the Stoke Mountain belt, bedded rocks already tightly folded were eroded before the overlying

Middle Ordovician beds were deposited. It seems logical to include these disturbances as a phase of the Taconic disturbance because the Middle Ordovician unconformities mark a lasting change from the deposit of limestone to the deposit of shale and coarser clastic rocks in the belt west of the Vermont-Quebec geanticline. Most notable are the clastics of the Late Ordovician exogeosyncline, which were derived from the Vermont-Quebec geanticline and vicinity as the orthogeosynclinal belt continued to rise while the climax of the main Taconic disturbance approached. (*See also* Rosenfeld, 1968, p. 196; Zen 1968, p. 130, 138.)

The Stoke Mountain geanticline also provided clastic sediments during the Middle and Late Ordovician. Subsidence of the adjoining geosynclinal belts, which possibly stopped during the Stoke Mountain phase of the Taconic disturbance, was resumed in the Middle Ordovician. Subsidence was not as great in a belt about 30 miles to the southeast of the Stoke Mountain geanticline, where as a consequence the Somerset geanticline appeared. The first appearance of the geanticline was in, or possibly sometime earlier than, the Middle Ordovician. Granitic rock (Highlandcroft Plutonic Series) was emplaced nearby in the late Middle or Late Ordovician (Pl. 3).

New generations of tight longitudinal folds, virtually inseparable from the old, formed in the newly deposited Middle Ordovician bedded rocks as geosynclinal subsidence and episodes of uplift continued with little change in the tectonic setting.[54] The relations of the late generations of early folds to lower Paleozoic plutonic rocks and to continually forming regional foliation remained essentially the same.

The early Paleozoic closed in most of the region with general uplift and erosion in the Late Ordovician to Early Silurian; this uplift and erosion marked the culmination of the Taconic disturbance. Southeast of the Somerset geanticline, however, geosynclinal subsidence continued from the Ordovician into the Silurian and little, if any, record of the Taconic disturbance is found.

(*See also* Boucot and others, 1964, p. 89–90; Green, 1968, p. 1632–1634; Pavlides and Berry, 1966, p. B60; Osberg and others, 1968, p. 244, 250; Pavlides and others, 1968; Zen, 1967, p. 64–69.)

MIDDLE PALEOZOIC

Biostratigraphic Chronology of the Middle Paleozoic

Quasi-cratonic middle Paleozoic strata eroded from the Champlain-St. Lawrence valley belt northwest of the Green Mountain-Sutton Mountain anticlinorium were probably Upper Silurian or higher, judged from the absence of Lower and Middle Silurian to the south near the middle course of the Hudson River. Presence of Lower Silurian rocks on Anticosti Island in the Gulf of St. Lawrence suggests, however, that some Lower and Middle Silurian units may have once entered the lower St. Lawrence valley. (*See* Berdan, 1964, Fig. 2; Boucot, 1968, Fig. 6–2, 6–3, 6–4, 6–5, 6–6; Cady, 1960, Pl. 3; Pavlides and others, 1968, p. 76.)

The biostratigraphically established middle Paleozoic—Silurian and Devonian

—bedded rocks southeast of the Green Mountain-Sutton Mountain anticlinorium include Early, Middle, and Late Silurian, and the Early and Middle Devonian. (*See* Boucot, 1961, p. 170; 1968, p. 92; Boucot and Arndt, 1960; Boucot and Drapeau, in press; Boucot and Johnson, 1967, p. 54–55; Boucot and Thompson, 1958, 1963; Boudette and others, 1967, p. 28–34, 46–49; Clark, 1923; 1937, p. 50–51; 1942; Doll and others, 1961; Harwood, 1968; Laverdière, 1936, p. 37–38; Naylor and Boucot, 1965, p. 159–162; St. Julien, 1963a, p. 179–180; *see also* Cady, 1960, p. 555.) These fossiliferous rocks are chiefly miogeosynclinal, or are of a lithofacies more nearly of that type interbedded in the eugeosynclinal zone. The eugeosynclinal zone is relatively unfossiliferous. (The geographic and stratigraphic limits of the rocks of Middle Devonian age are uncertain and are included with those designated lower Devonian on Plate 2 and Early Devonion on Plate 3.)

Distinctive stratigraphic units (including the Peasley Pond Conglomerate, Shaw Mountain Formation, and Clough Quartzite) are at the base of the Silurian in both the miogeosynclinal and eugeosynclinal zones on the east limb of the Green Mountain-Sutton Mountain anticlinorium and on the Bronson Hill-Boundary Mountain anticlinorium (Pl. 2). The various units contain quartz conglomerates and quartzites that are lithically and sequentially comparable over wide areas, and thus provide a regional key to the Lower and Middle Silurian chronology (Boucot and Thompson, 1963, p. 1316–1318; Cady, 1960, p. 551; 1968b, p. 159; Doll and others, 1961; Pavlides and others, 1968, p. 74–75).

The Devonian chronology in the intervening Connecticut Valley-Gaspé synclinorium is partly tied to the tracing of a distinctive intermittent metavolcanic unit; the Standing Pond Volcanic Member of the Waits River Formation and its correlatives (Doll and others, 1961), which extends principally parallel to the strike on the southeast limb of the synclinorium (Cady, 1960, p. 553). This unit, the various parts of which are considered nearly isochronous, pinches out laterally westward into interbedded crystalline limestones and slates (Waits River Formation) near the southeast edge of the miogeosynclinal zone and eastward into interbedded schist and quartzite (Gile Mountain and Littleton Formations). (*See* Doll and others, 1961; *also* Johansson, 1963, p. 72; Myers, 1964, p. 19–21.) At its northeast extremity this unit appears to correlate with the upper (mafic volcanic) member of the Kidderville Formation, the lower (felsic volcanic) member of which can be traced northeastward along strike into the fossiliferous Lower Devonian Seboomook Formation (Green, 1968, p. 1619, 1622; Green and Guidotti, 1968, p. 262; Hatch, 1963, p. 74–75; Myers, 1964, p. 27–28).

The Acadian orogeny is a late Middle Devonian or younger event, as shown by the biostratigraphic record along the southeast limb of the Green Mountain–Sutton Mountain anticlinorium in Quebec (Boucot, 1968, p. 92–93). Possibly two pulses of the Acadian orogeny are tied to the biostratigraphic record in the Connecticut Valley-Gaspé synclinorium in northeastern Maine, east of the region of the present study (Doyle and Hussey, 1967; Pavlides and others, 1964, p. C33, C37). There, various steeply dipping and closely folded and cleaved Silurian and Lower Devonian units are overlain unconformably by strata assigned to the late Middle Devonian (Mapleton Sandstone—Boucot, 1968,

p. 89) that are themselves gently flexed in an open syncline. An interesting problem concerns the question of whether the tight folds beneath the unconformity are of the early fold regime as their style suggests. If they are, the open syncline may very well be of the late regime, which time would suggest that the climactic pulse of the Acadian orogeny responsible for the interregional map pattern was in the late Middle Devonian or possibly later. (*See also* Boucot, Field, and others, 1964, p. 93–96; Boucot, Griscom, and Allingham, 1964, p. 1; Neuman and Rankin, 1966, p. 16; Poole, 1967, p. 33; Rankin, 1968, p. 366–367.) Alternatively, the tight folds are of the late regime—owing their style to the possibility that the rocks had not been deformed and thus made competent, in earlier regimes (R. H. Moench, oral commun., 1967).

Radiometric Chronology of the Middle Paleozoic Rocks

The K-Ar values from biostratigraphically dated middle Paleozoic rocks, like those from the lower Paleozoic, suggest mainly the times of metamorphism and subsequent cooling. Forty-one of these values from rocks at 36 widely distributed localities (Table 3, *see also* Faul and others, 1963, Fig. 2, p. 4, Fig. 3, p. 15; Goldsmith, 1964; Zartman and others, in press) range from 397 to 137 million, corresponding to a time span of more than 250 m.y. that includes the interval from the Early Devonian to Late Jurassic. The younger limits of this time span are 108 m.y. younger than the youngest K-Ar ages from lower Paleozoic rocks (Table 2) and 181 m.y. younger than those from Precambrian (Table 1). The greatest concentration of K-Ar dates, not only in middle Paleozoic rocks but in the Precambrian and lower and middle Paleozoic rocks as a whole (Tables 1, 2, and 3), is in the vicinity of 360 m.y.

The radiometric dates of Devonian metamorphism, deformation, and igneous intrusion are indicated most clearly near the northeast end of the Bronson Hill-Boundary Mountain anticlinorium in Maine (Pl. 1). In that area, slates containing Early Devonian fossils (Albee, 1961, p. C53) give K-Ar values (Table 3) of 350 ± 15, 368 ± 15, and 377 ± 18 that have led to the conclusion that the time of metamorphism of the slates was at least 360 m.y. ago (Hurley and others, 1959, p. 948, 949). These slates correlate stratigraphically with some of the closely folded and cleaved units that underlie the gently warped Middle Devonian Mapleton Sandstone discussed under "Biostratigraphic Chronology of the Middle Paleozoic;" this correlation indicates that the regional metamorphism and close folding, possibly the latest of the early regime and not necessarily of the late regime that is perhaps chiefly responsible for the map pattern, is at least as old as Middle Devonian. Moreover, these relations place the 360 m.y. date ascribed to the Acadian orogeny (Poole and others, 1964, p. 75, Fig. 5; *see also* Fairbairn and others, 1960, p. 405, Fig. 2; Faul and others, 1963, p. 2, Fig. 1; Long, 1962, p. 1004; Long and Kulp, 1962, p. 988–989) somewhere in the interval late Early to probably early Late Devonian.[55]

Discordant calc-alkalic granitic ("New England") plutons that intrude the Lower Devonian slates at the northeast end of the Bronson Hill-Boundary Mountain anticlinorium and also the zones of contact hornfels that adjoins these plutons, show comparable K-Ar ages (Table 3) of 360 ± m.y. (Faul and others, 1963, p. 8, 12; Hurley and others, 1959; Zartman and others, in press); but their assignment to the Acadian orogeny is uncertain inasmuch as none of the same type plutons have been found that intrude or are unconformably overlain by Middle Devonian rocks. The discordant contacts of such plutons in New Hampshire, Vermont, and Quebec make it clear, however, that they postdate the late fold regime marking the climax of the Acadian orogeny. They are thus of Middle or Late Devonian age, or both.[56] Moreover, these plutons are overlain unconformably in eastern Maine and New Brunswick by Mississippian strata (Page, 1968, p. 374).

The rocks showing Devonian metamorphic and intrusive dates (Pl. 1), without later overprint, are in a northeast-trending belt that extends beyond the confines of the region of the present study into northern Maine (Faul and others, 1963, p. 4, Fig. 2) and southeastern New York (Long, 1962, Fig. 1, p. 998; Long and Kulp, 1962, Fig. 2, p. 980). Northwest of this belt, in the Sutton, Notre Dame, and Shickshock Mountains of Quebec and in the lower Hudson River valley in New York, are the bedded rocks that were metamorphosed chiefly in the Middle Ordovician and earlier times (Lajoie and others, 1968, p. 621; Lespérance, 1959, p. 2)—back at least to the beginning of the Late Cambrian. Possibly, however, the Devonian metamorphism overstepped the lower Paleozoic metamorphism in west-central Vermont (Crosby, 1963, p. 123–124).

To the southeast, in New Hampshire, central Massachusetts, and eastern Connecticut, early and middle Paleozoic metamorphic and igneous activities are shown only by a few Rb-Sr whole-rock isochron and Pb-alpha measurements (Brookins and Hurley, 1965; Naylor and Wasserburg, 1966; Snyder, 1964, Table 2, p. 28; Zartman and others, 1965). Among these measurements, an age of as much as 355 $\pm$ 10 m.y. is shown for igneous activity in the middle Paleozoic in Connecticut. Late Paleozoic dates, discussed below, predominate (Pl. 1) and probably reflect chiefly delayed cooling of the middle Paleozoic metamorphic and igneous rocks. The middle Paleozoic age of certain igneous rocks with the late Paleozoic radiometric overprint is clear, however, from their structural and compositional similarities with the discordant calc-alkalic plutons northwest of the late Paleozoic overprint. Inasmuch as they are structurally undisturbed, the discordant calc-alkalic plutons indicate the minimum date of folding in areas both with and without the overprint.

Summary of Middle Paleozoic Tectonic Events

The northwestern part of the orthogeosyncline was stabilized to a quasi-cratonic belt at the beginning of the middle Paleozoic. Thus in the Early Silurian the miogeosynclinal zone overlapped southeastward on the eugeosynclinal zone southeast of the Vermont-Quebec geanticline (Pl. 3). The overlap extended southeastward across the Stoke Mountain geanticline and into the tract between the Stoke Mountain and the Somerset geanticlines, in the Late Silurian and Early Devonian.

Rapid geosynclinal subsidence resumed southeast of both the Stoke Mountain geanticline and the southern part of the Vermont-Quebec geanticline after a brief respite in the Early and Middle Silurian. During this time quartz conglomerates, quartz sandstones, and limestones were deposited and the Somerset geanticline stood still, partly above the sea. The geosynclinal tract between these geanticlines became the Silurian and Devonian belt of lateral transition southeastward from the miogeosynclinal to the eugeosynclinal zone. The Somerset geanticline still partly emergent continued to supply clastic sediments to the adjacent geosynclinal troughs. (*See* Green, 1968, p. 1634–1635; Naylor and Boucot, 1965.)

An intrageosynclinal uplift (Pl. 3 and Appendix A), from which sediments, recumbent folds, possibly slides, and also intrastratal diapiric folds moved northwest and possibly to the southeast, probably formed in the geosynclinal trough south of the Somerset geanticline in New Hampshire. The folds that face northwest show about the same attitude as do the northwest-facing folds formed in the same and similar settings exclusively during the early Paleozoic. They formed in the Early to Middle Devonian and therefore are ascribed to the Acadian orogeny although their style and genesis remained of the early fold regime.

Mafic and intermediate, and rarely, ultramafic intrusive rocks, continued to be emplaced in the eugeosynclinal zone during the middle Paloezoic. (*See* footnote 11, p. 31.) Concordant calc-alkalic granitic plutons (Bethlehem Gneiss and Kinsman Quartz Monzonite) were intruded in its southeastern belts. The regional foliation continued to form virtually parallel to that formed in the early Paleozoic and parallel to axial surfaces and limbs of new isoclinal folds, and it was superimposed upon the newly formed ultramafic to felsic intrusive rocks that appeared in the middle Paleozoic.

The middle Paleozoic was climaxed by growth of the domes and arches and concomitant evolution of the late longitudinal folds, thrust faults, and residual shear stresses that eventually produced joints. These events mark the height of the Acadian orogeny in the Devonian.[57] The discordant calc-alkalic plutons intruded both the bedded rocks and the concordant calc-alkalic plutons very soon after the Acadian climax, but before the end of the Devonian. They opened incipient joints transverse to bedding and foliation, and also entered zones of weakness parallel to them. With these events much of the orthogeosyncline had become cratonized.

The last events of the middle Paleozic history, events which probably continued into the Mississippian, were uplift and concomitant erosion of rocks deformed in the Acadian orogeny. During erosion the rocks (those now exposed at the ground surface) cooled, thus restarting the K-Ar systems about 360 m.y. ago. (*See also* Hadley, 1964.) Joints probably started to open with the erosional unloading, loss of confinement, and release of residual shear stress accumulated during folding. These events took place in a belt that coincides fairly closely with the Green Mountain-Sutton Mountain anticlinorium and Connecticut Valley-Gaspé synclinorium, and that encroaches especially upon the northern part of the Boundary Mountain anticlinorium. Meanwhile, the rocks to the southeast in and near the Merrimack synclinorium stood still or resumed subsidence, as they show little evidence of the critical cooling until the Permian.

LATE PALEOZOIC

A metamorphic overprint on middle Paleozoic and older rocks is the only record (Tables 2 and 3) of the late Paleozoic in the region of the present study (Pl. 1). The late Paleozoic epieugeosynclinal bedded rocks in southeastern New England, which lie unconformably on lower and middle Paleozoic granitic rocks (Bottino and others, 1968; Hurley and others, 1960a, p. 247–248) and are chiefly Pennsylvanian (Chute, 1966, p. 27–34; Quinn and Oliver, 1962), are absent, perhaps eroded, to the northwest. The overprint coincides fairly closely with rocks of sillimanite grade in and near the Merrimack synclinorium; it is found locally in lower grade rocks, northwest of the sillimanite terrane in the northern part of the Bronson Hill-Boundary Mountain anticlinorium (Zartman and others, in press). It is middle Permian, showing K-Ar ages of 240 to 260 m.y. (Table 3), measured on biotite (Faul and others, 1963, p. 14–15; Zartman and others, in press), and an Rb-Sr age of 245 ± 10 m.y. (Zartman and others, 1965, p. D5) on muscovite from nonfoliate pegmatite. These dates are probably close to the time of the final cooling event. An Rb-Sr isochron age of 255 ± 10 m.y. measured on biotite and plagioclase from gneiss (Monson Gneiss, Brookins, 1967) marks heating somewhat earlier in the Permian (*see also* Clark and Kulp, 1968, p. 885–886). All the Permian dates agree fairly closely with K-Ar metamorphic ages of 250 to 255 m.y. in the Pennsylvanian rocks (Hurley and others, 1960a, p. 252). The K-Ar ages of between 260 and 350 m.y. in the area of the present study, and in general west and northwest of the Permian overprint, probably in the main reflect earlier cooling in response to erosional unloading after the Acadian orogeny (*compare with* Clark and Kulp, 1968, p. 879).

Unmetamorphosed felsic volcanic rocks, the Moat Volcanics (Billings, 1956, p. 35–37) that lie unconformably on the metamorphic rocks mainly in eastern New Hampshire, are possibly of Permian age (Goldsmith, 1964; *see also* Williams and Billings, 1938, p. 1025–1026).[58] If these volcanic rocks are of Permian age the unconformity beneath them marks the Appalachian orogeny.

The meagerly recorded late Paleozoic events must be largely inferred. The rocks of the Merrimack synclinorium that show the Permian metamorphic overprint stood still or possibly resumed subsidence in the Mississippian. This stillstand or subsidence apparently continued into the middle Permian. Then the Lower Devonian and older rocks of sillimanite grade, now exposed at the ground surface in the Merrimack synclinorium, were uplifted, eroded, and critically cooled, resetting the K-Ar systems about 250 m.y. ago. The northwesterly drainage of northwestern New England and adjacent Quebec (Pl. 1) possibly started at this time; headwaters were in western New Hampshire and Maine.

If the rocks of the Merrimack synclinorium that show the Permian overprint subsided rather than stood still before uplift and erosion in the Permian, they probably formed the northwestern extremity of the epieugeosyncline whose Pennsylvanian terrestrial and shallow-water bedded rocks are preserved from erosion in southeastern New England. Regardless of their original extent to the northwest, the Pennsylvanian sediments were probably derived from the belt of Acadian folding and uplift to the northwest. They were in turn folded, metamorphosed, faulted, and uplifted, presumably in the Appalachian orogeny (Quinn and others, 1957, p. 552; Quinn and Oliver, 1962, p. 69–71),[59] which in this region culminated probably at about the time of uplift in the middle Permian.

MESOZOIC

The high-angle faults, some of which bound the Upper Triassic taphrogeosynclinal bedded rocks, are intersected by or transect discordant alkalic plutons and related dike rocks of Late Triassic or Early Jurassic (White Mountain Plutonic Series), possibly Late Jurassic or Early Cretaceous (Mt. Ascutney pluton), and Early Cretaceous (Monteregian plutons) age. The radiometric dates of the plutons and dike rocks are, respectively, about 180, 137 to 125, and 125 to 96 m.y. ago, as shown by K-Ar, U-Pb, Rb-Sr, and Rb-Sr whole-rock isochron measurements (Appendix F; Fairbairn and others, 1963; Faul and others, 1963, p. 16; Hurley and others, 1961a, p. 156; Tilton and others, 1957, p. 364; Zartman and others, 1967, p. 862, 867). The faults, therefore, are probably of Late Triassic to Early Cretaceous age. (*See also* Lyons and Faul, 1968, p. 312–313.)

The earliest Mesozoic event was probably continued widespread erosion of land areas uplifted in the belts both of the Acadian orogeny in northwestern New England and Quebec and of the Appalachian orogeny in southeastern New England and the Maritime Provinces. At least 12 miles of Paleozoic rock was eroded, according to estimates of thickness arrived at by reconstruction of recumbent early folds (30 km) and by calculation of lithostatic pressure and possibly tectonic overpressure necessary for metamorphism (20 km) (Robinson, 1966; 1967a, p. 40; *see also* footnote 19, p. 46). A surface of moderate relief soon formed in southern New England (Fritts, 1963, p. 278–279). This was followed in the Late Triassic by downfaulting between belts of Acadian domal anticlines in the Connecticut River region, and concomitant terrestrial sedimentation, mafic hypabyssal intrusion, and mafic volcanism of the taphrogeosynclinal trough. This trough probably diverted (pirated) the headwaters of any previously established northwesterly drainage. The high-angle faults along Lake Champlain and the St. Lawrence River probably formed somewhat later in the Mesozoic. The discordant alkalic plutons and related dike rocks were emplaced chiefly during the Jurassic and Early Cretaceous. Widespread selective downwasting (weathering and erosion) of variously resistant rocks in the Paleozoic and Mesozoic terranes continued during the late Mesozoic.[60] Most of the joints in the region had probably formed before the end of the Mesozoic, and therewith the greater part of erosional unloading, loss of confinement, and release of residual shear stress that had accumulated particularly during Acadian flexural folding, was completed.

CENOZOIC

Streams that flow northwestward across the Green, Sutton, and Notre Dame Mountains in New England and Quebec (Pl. 1) possibly survived from drainage that started on the northwest side of mountains raised in the Appalachian orogeny in latest Paleozoic, as already suggested. They flow nearly parallel to the systematic joints that transect the late longitudinal folds of the Green Mountain-Sutton Mountain anticlinorium, having partly adjusted to the joint pattern probably during the Cenozoic. Valleys and uplands selectively downwasted since the Mesozoic, a few unconsolidated Tertiary deposits, and Quaternary glacial features record the Cenozoic.

The lower Paleozoic miogeosynclinal and exogeosynclinal terranes in the Champlain and St. Lawrence Valleys are the most deeply weathered and eroded. The middle Paleozoic miogeosynclinal terrane southeast of the Green, Sutton, and Notre Dame Mountains also underlies relatively low areas, but is less dissected probably because carbonate rocks are less common. The low-lying lower Mesozoic taphrogeosynclinal terrane in the Connecticut Valley probably reflects less resistant bedrock than that of adjoining crystalline terranes and possibly downfaulting. The intervening areas, underlain locally by the Precambrian basement but chiefly by equally resistant predominantly metamorphic lower and middle Paleozoic eugeosynclinal terranes intruded by large bodies of granitic rocks, were left as uplands.[61]

Tertiary beds appear to have been deposited a little off the present New England coast (Uchupi, 1966, p. 3022–3024). Tertiary lignite and iron deposits lie upon Lower Cambrian dolomite bedrock along the east side of the miogeosynclinal terrane in the Champlain Valley and vicinity (Barghoorn and Spackman, 1949; Traverse and Barghoorn, 1953; Traverse, 1955). The lignite was evidently deposited inland and then let down into the valley by solution of the underlying carbonate rocks, whose insoluble residue including the iron was added to the unconsolidated surficial deposits. Recent work in a homologous part of the Great Valley of the Folded Appalachians (Hack, 1965, p. 63–66; Pierce, 1965, p. C154–C155) shows that this happened not only to Tertiary but to Cretaceous lignites.[62]

Finally, in the Quaternary the bedrock surface was depressed under the weight of continental glacier ice and then rebounded as the ice melted (MacClintock and Stewart, 1965, p. 10; McDonald, 1968; Stewart, 1961, p. 86–88). As denudation progresses and as a response to the removal of static load occurs, the rocks develop joints (most noticeably sheeting in granitic rocks) subparallel to the slopes of the ground surface. (*See* Jahns, 1943; *also* Hast, 1967.) Today, in quarries, the sudden release of the load on previously confined rock produces rock bursts (White, 1946).

FOOTNOTES

[43] The radiometric time scales used here are those of both Holmes (1959) and Kulp (1961), unless otherwise specified.

[44] The uppermost rocks were not the first eroded so do not yield the oldest dates, as discussed by Watson (1964, p. 457–458) concerning the Scottish Highlands.

[45] The terms *early*, *middle*, and *late* Precambrian as used here correspond respectively to Archean (>2.49 b.y.), early Proterozoic (Aphebian) (2.49–1.73 b.y.), and middle (Helikian) and late (Hadrynian) Proterozoic (1.73–0.57 b.y.) as employed by Stockwell (1963b, Fig. 3; 1964, Fig. 2; 1965a, and b; 1968, Table 1). The basis for this usage is discussed more fully by James (1958, p. 28–31).

[46] The term "Grenville orogeny" or "Grenvillian orogeny" is avoided because the name

"Grenville" as originally used applies to bedded rocks of the Grenville Series, which are probably much older than the 1-b.y. event. The "orogeny" is very likely a combination of one or more successive orogenic and related igneous events upon which the final metamorphic overprint, connected igneous features, and the billion-year event are superimposed. (*See also* Engel, 1963, p. 150; Gilluly, 1966, p. 105–109; Osborne and Morin, 1962, p. 135.) The traditional term "Grenville province," referring to the overprint, is kept purely for convenience, although it too is improper inasmuch as the overprint extends far beyond the terrane of the Grenville Series. (*See also* footnote 1, p. 14.) The age of 1530 m.y. was obtained from zircon that was probably formed in igneous rocks differentiated at that time, but whose sedimentary detritus became part of the upper Precambrian bedded rocks in which the zircon was found. (*See also* Fairbairn and others, 1967, p. 326.)

[47] The general problem of the dating of the base of the Cambrian in the light of which the present discussion must be read and interpreted, has recently been presented by Glaessner (1963a, and b). The special problem is nearly identical with that of the "late Precambrian"—Early Cambrian in Virginia described by Bloomer and Werner (1955, p. 598–599).

[48] Several units in northwestern Vermont (Mill River Conglomerate, Skeels Corners Slate, Rockledge Conglomerate, and Hungerford Slate Members of the Sweetsburg Formation and the Saxe Brook Dolomite—Doll and others, 1961) were tentatively assigned to the Middle Cambrian by Cady (1960, p. 554, footnote 14), following a similar assignment of these rocks by Shaw (1958, p. 548) to the *Cedaria* faunal zone, of which they are representative. They are here reassigned to the Upper Cambrian following recent designation by A. R. Palmer (oral commun., 1962; *see also* Robison, 1964; Shaw, 1966a, p. 275) of the base of the *Cedaria* zone as the base of the Upper Cambrian.

[49] A. R. Palmer (written commun., 1965) reported that he and Franco Rasetti had discovered Middle Cambrian fossils in rocks in the central Taconic region east of Albany and Troy, New York, that according to J. B. Thompson, Jr. (oral commun., 1966), correlate with part of the Hatch Hill Formation of the northern Taconic region. (*See also* Zen, 1967, p. 54.)

[50] The stratigraphic section near Lake Memphremagog in Quebec, traced north from the Moretown Member of the Missisquoi Formation in Vermont, includes upper parts of the Brompton Formation and unconformably above the Brompton includes all of the Magog Formation of St. Julien (1961b, p. 4–8; 1963b, p. 4–8; 1965). It is about the same section as that of the "Beauceville Group" and "Bolton Groups" as mapped by Cooke (1951) near the international boundary (Cady, 1960, p. 546; 1968b, p. 158; Cady and others, 1963, Table 1), and as that of the Magog slates and metabasalt as shown by Clark (1934, p. 4, Fig. 2, p. 12). Its facies, volcanic and pelitic, are most nearly like those of the Cram Hill Member of the Missisquoi Formation, which occurs discontinuously in the uppermost Missisquoi, where it grades laterally into upper strata of the Moretown Member (Doll and others, 1961). The pelitic facies of the Cram Hill are lithologically like those of the Magog Formation of St. Julien (1961b, p. 8; 1963b, p. 408).

[51] The radiometric ages of the felsic dikes that cut the ultramafic rocks, which are themselves emplaced in strata beneath correlatives of the fossiliferous Middle Ordovician southeast of the Green Mountain-Sutton Mountain anticlinorium in Quebec, have been used by Lamarche (1965, p. 149) to date the units beneath the fossiliferous rocks as Early Ordovician or earlier. It seems likely, however, in view of the equivocal relationships of the felsic dikes, that some of the rocks beneath the fossiliferous Middle Ordovician could also be of Middle Ordovician age.

[52] The Highlandcroft Plutonic Series has been given a Pb/alpha age (Larsen method) of 385 ± 27 m.y. (Lyons and others, 1957, p. 529–530; *see also* Cady, 1960, p. 563). This age would incorrectly place the Highlandcroft in the Middle Devonian of the revised geologic time scales (Holmes, 1959, p. 204; Kulp, 1961, p. 1111), although this date did fall in the Ordovician of a previously accepted time scale (Holmes, 1947, p. 123). Additional measurements made by the most recent Pb/alpha methods (Rose and Stern, 1960; Stern and Rose, 1961), or Rb-Sr whole-rock isochron analysis, appear necessary. (*See also* Naylor, 1969, p. 414–415.)

[53] The Stoke Mountain phase of the Taconic disturbance approximates "Phase 1" of the "Taconic orogeny" of Rickard (1965, p. 531). The "early phase of the Taconic orogeny" of Poole (1967, p. 20) refers to epeirogenic movements of the craton.

[54] The post-Middle Ordovician early folds approximate the folds (but not the thrusts) of "Phase 2" of the "Taconic orogeny" of Rickard (1965, p. 531–533), and some of the folds (but perhaps not the slides) of the middle phase of the "Taconian orogeny" of Poole (1967, p. 21–23). (*See also* footnote 57, p. 118.)

[55] The geologic time scale of Kulp (1961, p. 1111) places the 360 m.y. radiometric date in the Late Devonian of the biostratigraphic age column. The relationships discussed here show that

the time scale must be revised to place the 360 m.y. date in correspondence probably with the Middle Devonian part of the biostratigraphic column, nearly as recently diagrammed by Poole, Kelley, and Neale (1964, Fig. 5), and as according to "the Geological Society Phanerozoic time scale 1964" (Harland and others, 1964, p. 260–262).

[56] The Lower-Middle Devonian boundary is probably something less than 390 m.y. old and the Devonian-Mississippian boundary is 345 ± 10 m.y. old (Kulp, 1961, p. 1110; Smith and others, 1961, p. 173).

[57] The Devonian late longitudinal folds and thrust faults are, respectively, the folds of "Phase 3" and the thrusts of "Phase 2" of the "Taconic orogeny" of Rickard (1965, p. 531–535). (*See* footnote 54, p. 117.) They include folds and thrusts of the "late phase" of the "Taconic orogeny" and the "main phase" of the "Acadian orogeny" of Poole (1967, p. 24–26, 33–77). Assignment of the late longitudinal folds and the thrust faults to the Taconic disturbance is based on the assumption that K-Ar dates on micas that are not bent but form polygonal arches in minor folds and some of which parallel the axial plane cleavage of the folds, are true dates (Rickard, 1965, p. 526–531). The author believes that the critical K-Ar dates are hybrid dates for reasons discussed in Appendix G. (*See also* Dennis, 1968.)

[58] A Permian age of the Moat Volcanics is possibly valid, although it was assigned for reasons that turn out to be variously questionable and incorrect. The Moat has been considered comagmatic with the White Mountain plutons, because its exposures are closely associated with the plutons in down-dropped blocks in ring-dike complexes and because of compositional affinities (Billings, 1956, p. 130–131; Noble and Billings, 1967). The plutons were assigned a Permian age of 234 ± 22 m.y., according to a now obsolete method of Pb/alpha measurements (Lyons and others, 1957, p. 538–542; *see also* Rose and Stern, 1960, and Stern and Rose, 1961). They are now found by K-Ar and Rb-Sr measurements to be 170 to 185 m.y. old (Hurley and others, 1960a, p. 254; Tilton and others, 1957, p. 364)—Early Jurassic or Late Triassic on the most recent time scales (Holmes, 1959, p. 204; Kulp, 1961, p. 1111).

[59] The inclusive term "Appalachian orogeny" is preferred here for a possible orogenic phase because an alternate term "Alleghany orogeny" proposed by Woodward (1957, p. 1440) and employed by Faul and others (1963, p. 14) suggests instead of a Permian event one closely related in time to the Pennsylvanian Allegheny Group (Carswell and Bennett, 1963, p. 10–12, 37–46). Note the two spellings—Alleghany and *Allegheny*. A new term, "Maritime Disturbance" (Poole, 1967, p. 33, 40–41) will possibly prove most appropriate in New England as well as in adjacent parts of Canada.

[60] Discontinuous and nonselective downwasting—peneplanation—has long been postulated for the late Mesozoic and the Cenozoic of the Appalachian Mountains. But in the absence of clear evidence for peneplanes, especially in New England, the concept is here reconsidered in line with the arguments of Hack (1960, 1965, p. 3–10, 63–66) for continuous and selective downwasting. The "Fall Zone peneplane" was believed to have been formed, tilted, and its destruction commenced in a later ("Schooley") cycle of peneplanation, all during the late Mesozoic. (*See* Johnson, 1931, p. 49–50; Sharp, 1929, p. 35–36.)

[61] A widely concordant upland surface in southern New England has been referred to as the "Schooley peneplane," and discordant summits in northern New England have been considered to be monadnocks that rise above this peneplane. (*See* Atwood, 1940, p. 80–82; Johnson, 1931, p. 11; Sharp, 1929, p. 36.) The concordant summits may be explained, however, by uniform downwasting (Hack, 1960, p. 85–87) and the monadnocks (of which Mount Monadnock in New Hampshire is the type example) by the slower downwasting of more resistant rocks. Downwasting calculated at a rate of about 0.08 foot in 1000 years (that of a humid region, Schumm, 1963, p. H12), which followed faulting in the Late Triassic (180+ m.y. ago), is 14,400 feet. Inasmuch as altitudes greater than 14,400 feet appear reasonable at the onset of downwasting, it is unnecessary to invoke isostatic compensation to maintain a terrain above sea level, although such compensation possibly occurred.

[62] The Tertiary lignite in both the Great Valley of the Appalachians and in the Champlain Valley have until recently been interpreted as deposits laid down on floors of valleys cut in the uplifted "Schooley" base-leveled surface entirely after the valleys were formed (Barghoorn and Spackman, 1949, p. 37–38; King, 1950, p. 57–58). These valley floors, or the benches that are their relics left by more recent erosion, have been referred to the "Harrisburg peneplane" (Johnson, 1931, p. 21).

References Cited

Albee, A. L., 1957, Hyde Park quadrangle, Vermont—bedrock geology: U.S. Geol. Survey Geol. Quad. Map GQ-102.

——1961, Boundary Mountain anticlinorium, west-central Maine and northern New Hampshire, *in* Geological Survey Research, 1961: U.S. Geol. Survey Prof. Paper 424-C, p. C51-C54.

——1965, Phase equilibria in three assemblages of kyanite-zone pelitic schists, Lincoln Mountain quadrangle, central Vermont, *with analyses by* E. Bingham, A. A. Chodos, and A. D. Maynes: Jour. Petrology, v. 6, no. 2, p. 246–301.

——1968, Metamorphic zones in northern Vermont, p. 329–341, *in* Zen, E-an, White, W. S., Hadley, J. J., and Thompson, J. B., Jr., *Editors*, Studies of Appalachian Geology: Northern and Maritime: New York, Interscience Publishers, 475 p.

Althaus, Egon, 1967, The triple point andalusite-sillimanite-kyanite: Beitr. Mineralogie u. Petrographie, v. 16, no. 1, p. 29–44.

Ambrose, J. W., 1942, Preliminary map, Mansonville, Quebec: Canada Geol. Survey Paper 42–1, geol. map.

——1943, Preliminary map, Stanstead and Brome Counties, Quebec: Canada Geol. Survey Paper 43–12, geol. map.

——1957, The age of the Bolton lavas, Memphremagog district, Quebec: Naturaliste Canadien, v. 84, nos. 8–9, p. 161–170.

Aniruddha, De, 1968, Post-ultramafic dikes and associated rocks in the asbestos mining areas in the Eastern Townships, Quebec, *in* Report of activities, 1967, Part A, Geol. Survey Canada Paper 68–1, p. 108.

Appleyard, E. C., 1965, The Grenville Province and its tectonic significance—a discussion: Geol. Assoc. Canada Proc., v. 16, p. 51–57.

Atwood, W. W., 1940, The physiographic provinces of North America: Boston, Ginn and Co., 536 p.

Bailey, E. H., Irwin, W. P., and Jones, D. L., 1964, Franciscan and related rocks, and their significance in the geology of western California: California Div. Mines and Geology Bull. 183, 177 p.

Bain, G. W., 1931, Flowage folding: Am. Jour. Sci., 5th Ser., v. 22, p. 503–530.

Balk, Robert, 1927, A contribution to the structural relations of the granitic intrusions of Bethel, Barre, and Woodbury, Vermont: Vermont State Geologist 15th Biennial Rept. 1925–26, p. 39–96.

——1952, Fabric of quartzites near thrust faults: Jour. Geology, v. 60, no. 5, p. 415–435.

Balk, Robert, 1956a, Massachusetts portion of the Bernardston quadrangle, Massachusetts-Vermont—bedrock geology: U.S. Geol. Survey Geol. Quad. Map GQ-90.

——1956b, Massachusetts portion of the Northfield quadrangle, Massachusetts, New Hampshire, and Vermont—bedrock geology: U.S. Geol. Survey Geol. Quad. Map GQ-92.

Barghoorn, E. S., Jr., and Spackman, William, Jr., 1949, A preliminary study of the flora of the Brandon lignite [Vermont]: Am. Jour. Sci., v. 247, no. 1, p. 33–39.

Bateman, P. C., and Eaton, J. P., 1967, Sierra Nevada batholith: Science, v. 158, no. 3807, p. 1407–1417.

Bean, R. J., 1953, Relation of gravity anomalies to the geology of central Vermont and New Hampshire: Geol. Soc. America Bull., v. 64, no. 5, p. 509–537.

Becker, G. F., 1895, The torsional theory of joints: Am. Inst. Mining Engineers Trans., v. 24, p. 130–138, 863–867.

Béland, J. R., 1957, St. Magloire and Rosaire-St. Pamphile areas, electoral districts of Dorchester, Bellechasse, Montmagny, and L'Islet [Quebec]: Quebec Dept. Mines, Geol. Survey Br., Geol. Rept. 76, 49 p.

——1962, Ste. Perpétue area, Kamouraska and L'Islet Counties [Quebec]: Quebec Dept. Nat. Resources, Geol. Survey Br., Geol. Rept. 98, 22 p.

Berdan, J. M., 1964, The Helderberg Group and the position of the Silurian-Devonian boundary in North America: U.S. Geol. Survey Bull. 1180-B, p. B1-B19.

Berry, W. B. N., 1962, On the Magog, Quebec, graptolites: Am. Jour. Sci., v. 260, no. 2, p. 142–148.

——1968, Ordovician paleogeography of New England and adjacent areas based on graptolites, p. 23–34, *in* Zen, E-an, White, W. S., Hadley, J. J., and Thompson, J. B., Jr., *Editors*, Studies of Appalachian Geology: Northern and Maritime: New York, Interscience Publishers, 475 p.

Billings, M. P., 1933, Thrusting younger rocks over older: Am. Jour. Sci., 5th ser., v. 25, no. 146, p. 140–165.

——1937, Regional metamorphism of the Littleton-Moosilauke area, New Hampshire: Geol. Soc. America Bull., v. 48, no. 4, p. 463–565.

——1955, Geologic map of New Hampshire: U.S. Geol. Survey.

——1956, Bedrock geology, Part 2 *of* The geology of New Hampshire: New Hampshire Plan. and Devel. Comm., 203 p.

Billings, M. P., and Keevil, N. B., 1946, Petrography and radioactivity of four Paleozoic magma series in New Hampshire: Geol. Soc. America Bull., v. 57, no. 9, p. 797–828.

Billings, M. P., and White, W. S., 1950, Metamorphosed mafic dikes of the Woodsville quadrangle, Vermont and New Hampshire: Am. Mineralogist, v. 35, nos. 9–10, p. 629–643.

Billings, M. P., and Wilson, J. R., 1964, Chemical analyses of rocks and rock minerals from New Hampshire: New Hampshire Dept. Resources and Econ. Devel. Mineral Resources Survey, Part 19, 104 p.

Bird, J. M., and Rasetti, Franco, 1968, Lower, Middle, and Upper Cambrian faunas in the Taconic sequence of eastern New York: Stratigraphic and biostratigraphic significance: Geol. Soc. America Spec. Paper 113, 66 p.

Bird, J. M., and Theokritoff, George, 1967, Mode of occurrence of fossils in the Taconic allochthon (Abs.), *in* Abstracts for 1966: Geol. Soc. America Spec. Paper 101, p. 248.

Bloomer, R. O., and Werner, H. J., 1955, Geology of the Blue Ridge region in central Virginia: Geol. Soc. America Bull., v. 66, no. 5, p. 597–606.

Bottino, M. L., and Fullagar, P. D., 1966, Whole-rock rubidium-strontium age of the Silurian-Devonian boundary in northeastern North America: Geol. Soc. America Bull., v. 77, no. 10, p. 1167–1175.

Bottino, M. L., Fullagar, P. D., Fairbairn, H. W., Pinson, W. H., Jr., and Hurley, P. M., 1969, Rubidium-strontium study of the Blue Hills igneous complex and the Wamsutta Formation rhyolite, Massachusetts (Abs.), *in* Abstracts for 1968: Geol. Soc. America Spec. Paper 121, in press.

Boucot, A. J., 1961, Stratigraphy of the Moose River synclinorium, Maine: U.S. Geol. Survey Bull. 1111-E, p. 153–188.

——1968, Silurian and Devonian of the northern Appalachians, p. 83–94, *in* Zen, E-an, White, W. S., Hadley, J. J., and Thompson, J. B., Jr., *Editors*, Studies of Appalachian Geology: Northern and Maritime: New York, Interscience Publishers, 475 p.

Boucot, A. J., and Arndt, Robert, 1960, Fossils of the Littleton Formation (Lower Devonian) of New Hampshire: U.S. Geol. Survey Prof. Paper 334-B, p. 41–51.

Boucot, A. J., and Drapeau, Georges, (in press), The Siluro-Devonian rocks of Lake Memphremagog and their correlatives in the Eastern Townships, Quebec: Quebec Dept. Nat. Resources Spec. Paper 1.

Boucot, A. J., Field, M. T., Fletcher, Raymond, Forbes, W. H., Naylor, R. S., and Pavlides, Louis, 1964, Reconnaissance bedrock geology of the Presque Isle quadrangle, Maine: Maine Geol. Survey Quad. Mapping Ser., no. 2, 123 p.

Boucot, A. J., Griscom, Andrew, and Allingham, J. W., 1964, Geologic and aeromagnetic map of northern Maine: U.S. Geol. Survey Geophys. Inv. Map GP-312, text, 7 p.

Boucot, A. J., and Johnson, J. G., 1967, Paleogeography and correlation of Appalachian Province Lower Devonian sedimentary rocks: Tulsa Geol. Soc. Digest, v. 35, p. 35–87.

Boucot, A. J., and Thompson, J. B., Jr., 1958, Late Lower Silurian fossils from sillimanite zone near Claremont, New Hampshire: Science, v. 128, no. 3320, p. 362–363.

Boucot, A. J., and Thompson, J. B., Jr., 1963, Metamorphosed Silurian brachiopods from New Hampshire: Geol. Soc. America Bull., v. 74, no. 11, p. 1313–1333.

Boudette, E. L., Hatch, N. L., Jr., and Harwood, D. S., 1967, Geology of the upper St. John and Allagash River Basins, Maine: U.S. Geol. Survey open-file rept., 77 p.

Brace, W. F., 1953, The geology of the Rutland area, Vermont: Vermont Geol. Survey Bull. 6, 120 p.

——1958, Interaction of basement and mantle during folding near Rutland, Vermont: Am. Jour. Sci., v. 256, no. 4, p. 241–256.

Brookins, D. G., 1967, Rb-Sr age evidence for Permian metamorphism of the Monson Gneiss, west-central Massachusetts: Geochim. et Cosmochim. Acta, v. 31, no. 2, p. 281–283.

——1968, Rb-Sr age of the Ammonoosuc Volcanics, New England: Am. Jour. Sci., v. 266, no. 7, p. 605–608.

Brookins, D. G., and Hurley, P. M., 1965, Rb-Sr geochronological investigations in the Middle Haddam and Glastonbury quadrangles, eastern Connecticut: Am. Jour. Sci., v. 263, no. 1, p. 1–16.

Bucher, W. H., 1920–1921, The mechanical interpretation of joints: Jour. Geology, v. 28, no. 8, p. 707–730; v. 29, no. 1, p. 1–28.

Buddington, A. F., 1939, Adirondack igneous rocks and their metamorphism: Geol. Soc. America Mem. 7, 354 p.

Buddington, A. F., 1960, The origin of anorthosite re-evaluated: India Geol. Survey Recs., v. 86, Part 3, p. 421–432.

Buddington, A. F., and Whitcomb, Lawrence, 1941, Geology of the Willsboro quadrangle, New York: New York State Mus. Bull. 325, 137 p.

Cady, W. M., 1937, Middlebury synclinorium in west-central Vermont (Abs.): Geol. Soc. America Proc. 1936, p. 67.

——1945, Stratigraphy and structure of west-central Vermont: Geol. Soc. America Bull., v. 56, no. 5, p. 515–587.

——1950a, Fossil cup corals from the metamorphic rocks of central Vermont: Am. Jour. Sci., v. 248, no. 7, p. 488–497.

——1950b, Classification of geotectonic elements: Am. Geophys. Union Trans., v. 31, no. 5, p. 780–785.

——1956, Montpelier quadrangle, Vermont—bedrock geology: U.S. Geol. Survey Geol. Quad. Map GQ-79.

——1960, Stratigraphic and geotectonic relationships in northern Vermont and southern Quebec: Geol. Soc. America Bull., v. 71, no. 5, p. 531–576.

——1967, Geosynclinal setting of the Appalachian Mountains in southeastern Quebec and northwestern New England: Royal Soc. Canada Spec. Pub. 10, p. 57–68.

——1968a, Tectonic setting and mechanism of the Taconic slide: Am. Jour. Sci., v. 266, no. 7, p. 563–578.

——1968b, The lateral transition from the miogeosynclinal to the eugeosynclinal zone in northwestern New England and adjacent Quebec, p. 151–161, *in* Zen, E-an, White, W. S., Hadley, J. J., and Thompson, J. B., Jr., *Editors*, Studies of Appalachian Geology: Northern and Maritime: New York, Interscience Publishers, 475 p.

Cady, W. M., Albee, A. L., and Chidester, A. H., 1963, Bedrock geology and asbestos deposits of the upper Missisquoi Valley and vicinity, Vermont: U.S. Geol. Survey Bull. 1122-B, 78 p.

Cady, W. M., Albee, A. L., and Murphy, J. F., 1962, Lincoln Mountain quadrangle, Vermont—bedrock geology: U.S. Geol. Survey Geol. Quad. Map GQ-164.

Cady, W. M., and Chidester, A. H., 1957, Magmatic relationships in northern Vermont and southern Quebec (Abs.): Geol. Soc. America Bull., v. 68, no. 12, Part 2, p. 1705.

Cady, W. M., and Zen, E-an, 1960, Stratigraphic relationships of the Lower Ordovician Chipman Formation in west-central Vermont: Am. Jour. Sci., v. 258, no. 10, p. 728–739.

Carey, S. W., 1962, Folding: Alberta Soc. Petroleum Geologists Jour., v. 10, no. 3, p. 95–144.

Carswell, L. D., and Bennett, G. D., 1963, Geology and hydrology of the Neshannock quadrangle, Mercer and Lawrence Counties, Pennsylvania: Pennsylvania 4th ser., Geol. Survey, Ground Water Rept. W15 (Bull. W15), 90 p.

Chang, Ping Hsi, Ern, E. H., Jr., and Thompson, J. B., Jr., 1965, Bedrock geology of the Woodstock quadrangle, Vermont: Vermont Geol. Survey Bull. 29, 65 p.

Chapman, C. A., 1939, Geology of the Mascoma quadrangle, New Hampshire: Geol. Soc. America Bull., v. 50, no. 1, p. 127–180.

——1952, Structure and petrology of the Sunapee quadrangle, New Hampshire: Geol. Soc. America Bull., v. 63, no. 4, p. 381–425.

——1963, Structural control on magmatic central complexes of central New England (Abs.), *in* Abstracts for 1962: Geol. Soc. America Spec. Paper 73, p. 128.

Chapman, C. A., Billings, M. P., and Chapman, R. W., 1944, Petrology and structure of the Oliverian magma series in the Mount Washington quadrangle, New Hampshire: Geol. Soc. America Bull., v. 55, no. 4, p. 497–516.

Chapman, R. W., 1942, Ring structure of the Pliny region, New Hampshire: Geol. Soc. America Bull., v. 53, no. 10, p. 1533–1567.

Chapman, R. W., and Williams, C. R., 1935, Evolution of the White Mountain magma series: Am. Mineralogist, v. 20, no. 7, p. 502–530.

Chidester, A. H., 1953, Geology of the talc deposits, Sterling Pond area, Stowe, Vermont: U.S. Geol. Survey Mineral Inv. Field Studies Map MF-11.

——1962, Petrology and geochemistry of selected talc-bearing ultramafic rocks and adjacent country rocks in north-central Vermont: U.S. Geol. Survey Prof. Paper 345, 207 p.

——1968, Evolution of the ultramafic complexes of northwestern New England, p. 343–354, *in* Zen, E-an, White, W. S., Hadley, J. J., and Thompson, J. B., Jr., *Editors*, Studies of Appalachian Geology: Northern and Maritime: New York, Interscience Publishers, 475 p.

Chidester, A. H., Billings, M. P., and Cady, W. M., 1951, Talc investigations in Vermont, preliminary report: U.S. Geol. Survey Circ. 95, 33 p.

Chinner, G. A., 1966, The distribution of pressure and temperature during Dalradian metamorphism: Geol. Soc. London Quart. Jour., v. 122, Part 2, p. 159–186.

Christensen, M. N., 1963, Structural analysis of Hoosac nappe in northwestern Massachusetts: Am. Jour. Sci., v. 261, no. 2, p. 97–107.

Christman, R. A., 1959, Geology of the Mount Mansfield quadrangle, Vermont: Vermont Geol. Survey Bull. 12, 75 p.

Christman, R. A., and Secor, D. T., Jr., 1961, Geology of the Camels Hump quadrangle, Vermont: Vermont Geol. Survey Bull. 15, 70 p.

Church, W. R., 1968, Isotopic ages from the Appalachians and their tectonic significance: Discussion: Canadian Jour. Earth Sciences, v. 5, no. 5, p. 1331–1333.

Chute, N. E., 1966, Geology of the Norwood quadrangle, Norfolk and Suffolk Counties, Massachusetts: U.S. Geol. Survey Bull. 1163-B, 78 p.

Clark, G. S., and Kulp, J. L., 1968, Isotopic age study of metamorphism and intrusion in western Connecticut and southeastern New York: Am. Jour. Sci., v. 266, no. 10, p. 865–894.

Clark, S. P., Jr., Robertson, E. C., and Birch, Francis, 1957, Experimental determination of kyanite-sillimanite equilibrium relations at high temperatures and pressures: Am. Jour. Sci., v. 255, no. 9, 628 640.

Clark, T. H., 1921, A review of the evidence for the Taconic revolution: Boston Soc. Nat. History Proc., v. 36, no. 3, p. 135–163.

——1923, The Devonian limestone at St. George, Quebec: Jour. Geology, v. 31, no. 3, p. 217–225.

——1934, Structure and stratigraphy of southern Quebec: Geol. Soc. America Bull., v. 45, no. 1, p. 1–20.

——1936, A Lower Cambrian series from southern Quebec: Royal Canadian Inst. Trans., v. 21, Part 1, p. 135–151.

——1937, Lake Aylmer series, *in* Cooke, H. C., Thetford, Disraeli, and eastern half of Warwick map-areas, Quebec: Canada Geol. Survey Mem. 211, p. 43–52.

——1942, Helderberg faunas from the eastern townships of Quebec: Royal Soc. Canada Trans., 3rd Ser., v. 36, Sec. 4, p. 11–36.

Clark, T. H., 1947, Summary report on the St. Lawrence Lowlands south of the St. Lawrence River [Quebec]: Quebec Dept. Mines, Geol. Survey Br., Prelim. Rept. 204, 18 p., and Map 642.

——1951, New light on Logan's Line [Quebec]: Royal Soc. Canada Trans., 3rd Ser., v. 45, Sec. 4, p. 11–22.

——1954a, St. Jean, Quebec [geology and topography]: Quebec Dept. Mines Map 847.

——1954b, Beloeil, Quebec [geology and topography]: Quebec Dept. Mines Map 848.

——1955, St. Jean-Beloeil area [Quebec]—Iberville, St. Jean, Napierville-Laprairie, Rouville, Chambly, St. Hyacinthe, and Verchères Counties: Quebec Dept. Mines, Geol. Survey Br., Geol. Rept. 66, 83 p.

——1964a, Upton area, Bagot, Drummond, Richelieu, St. Hyacinthe, and Yamaska Counties [Quebec]: Quebec Dept. Nat. Resources Geol. Rept. 100, 37 p.

——1964b, St. Hyacinthe area (west half), Bagot, St. Hyacinthe, and Shefford Counties [Quebec]: Quebec Dept. Nat. Resources Geol. Rept. 101, 128 p.

——1964c, Yamaska-Aston area, Nicolet, Yamaska, Berthier, Richelieu, and Drummond Counties [Quebec]: Quebec Dept. Nat. Resources Geol. Rept. 102, 192 p.

——1966, Chateauguay area, Chateauguay, Huntingdon, Beauharnois, Napierville, and St. Jean Counties [Quebec]: Quebec Dept. Nat. Resources Geol. Rept. 122, 63 p.

Clark, T. H., and Eakins, P. R., 1968, The stratigraphy and structure of the Sutton area of southern Quebec, p. 163–173, *in* Zen, E-an, White, W. S., Hadley, J. J., and Thompson, J. B., Jr., *Editors*, Studies of Appalachian Geology: Northern and Maritime: New York, Interscience Publishers, 475 p.

Cooke, H. C., 1937, Thetford, Disraeli, and eastern half of Warwick map-areas, Quebec: Canada Geol. Survey Mem. 211, 160 p.

——1950, Geology of a southwestern part of the eastern townships of Quebec: Canada Geol. Survey Mem. 257, 142 p.

——1951, [Geological map] Magog-Weedon, Quebec: Canada Geol. Survey Map 994A.

Crosby, G. W., 1963, Structural evolution of the Middlebury synclinorium, west-central Vermont: Columbia Univ., New York, Ph.D. dissert., 136 p.

——1964, Structural evolution of the Middlebury synclinorium, west-central Vermont (Abs.), *in* Abstracts for 1963: Geol. Soc. America Spec. Paper 76, p. 37–38.

Currier, L. W., and Jahns, R. H., 1941, Ordovician stratigraphy of central Vermont: Geol. Soc. America Bull., v. 52, no. 9, p. 1487–1512.

Daly, R. A., 1903, Geology of Ascutney Mountain, Vermont: U.S. Geol. Survey Bull. 209, 122 p.

Davis, G. L., Hart, S. R., Aldrich, L. T., and Krogh, T. E., 1967, Geochronology, *in* Carnegie Inst. Washington Yearbook, Director Geophys. Lab. Ann. Rept. 1965–1966, p. 199–438.

Dennis, J. G., 1956, The geology of the Lyndonville area, Vermont: Vermont Geol. Survey Bull. 8, 98 p.

——1960, Zum Gebirgsbau der nördlichen Appalachen: Geol. Rundschau, v. 50, p. 554–577.

——1964, The geology of the Enosburg area, Vermont: Vermont Geol. Survey Bull. 23, 56 p.

——1968, Isotopic ages from the Appalachians and their tectonic significance: Discussion: Canadian Jour. Earth Sci., v. 5, no. 4, p. 959–961.

de Römer, H. S., 1958, Preliminary report on Lake Orford area, electoral districts of Brome, Shefford, and Sherbrooke [Quebec]: Quebec Dept. Mines, Mineral Deposits Br., Prelim. Rept. 372, 11 p.

de Römer, H. S., 1961, Structural elements in southeastern Quebec, northwestern Appalachians, Canada: Geol. Rundschau, v. 51, no. 1, p. 268–280.

——1963, Differentiation in Chagnon-Orford-Intrusiv-Komplex, Südostteil der Provinz Quebec, Kanada: Geol. Rundschau, v. 52, no. 2, p. 825–835.

de Sitter, L. U., 1964, Structural geology: New York, McGraw-Hill Book Co., 551 p.

de Waard, D., and Walton, M. S., 1967, Precambrian geology of the Adirondack highlands, a reinterpretation: Geol. Rundschau, v. 56, no. 2, p. 596–629.

Diment, W. H., 1953, A regional gravity survey in Vermont, western New Hampshire, and eastern New York: Harvard Univ., Cambridge, Ph.D. dissert., 176 p.

——1956, Regional gravity survey in Vermont, western Massachusetts, and eastern New York (Abs.): Geol. Soc. America Bull., v. 67, no. 12, Part 2, p. 1688.

——1964, Gravity and magnetic anomalies in northeastern New York (Abs.), *in* Abstracts for 1963: Geol. Soc. America Spec. Paper 76, p. 45.

——1968, Gravity anomalies in northwestern New England, p. 399–413 *in* Zen, E-an, White, W. S., Hadley, J. J., and Thompson, J. B., Jr., *Editors*, Studies of Appalachian Geology: Northern and Maritime: New York, Interscience Publishers, 475 p.

Dixon, H. R., and Lundgren, L. W., Jr., 1968, Structure of eastern Connecticut, p. 219–229, *in* Zen, E-an, White, W. S., Hadley, J. J., and Thompson, J. B., Jr., *Editors*, Studies of Appalachian Geology: Northern and Maritime: New York, Interscience Publishers, 475 p.

Dixon, H. R., Lundgren, L. W., Jr., Snyder, G. L., and Eaton, Gordon, 1963, Colchester nappe of eastern Connecticut (Abs.), *in* Abstracts for 1962: Geol. Soc. America Spec. Paper 73, p. 139.

Dixon, H. R., and Shaw, C. E., Jr., 1965, Geologic map of the Scotland quadrangle, Connecticut: U.S. Geol. Survey Geol. Quad. Map 392.

Doll, C. G., 1951, Geology of the Memphremagog quadrangle and southeastern portion of the Irasburg quadrangle, Vermont: Vermont Geol. Survey Bull. 3, 113 p.

Doll, C. G., Cady, W. M., Thompson, J. B., Jr., and Billings, M. P., 1961, Centennial geologic map of Vermont: Vermont Geol. Survey.

——1963, Reply to Zen's discussion of the Centennial geologic map of Vermont: Am. Jour. Sci., v. 261, no. 1, p. 94–96.

Donath, F. A., 1963, Fundamental problems in dynamic structural geology, *in* Donnelly, T. W., *Editor*, The earth sciences; problems and progress in current research: Chicago, Univ. Chicago Press, p. 83–103.

Donath, F. A., and Parker, R. B., 1964, Folds and folding: Geol. Soc. America Bull., v. 75, no. 1, p. 45–62.

Doyle, R. G., and Hussey, A. M., 2d, 1967, Preliminary geologic map of Maine: Maine Geol. Survey.

Dresser, J. A., 1912, Reconnaissance along the National Transcontinental Railway in southern Quebec: Canada Geol. Survey Mem. 35, 42 p.

——1913, Preliminary report on the serpentine and associated rocks of southern Quebec: Canada Geol. Survey Mem. 22, 103 p.

Dresser, J. A., and Denis, T. C., 1944, Descriptive geology, v. 2 *of* Geology of Quebec: Quebec Dept. Mines Geol. Rept. 20, 544 p.

Duquette, Gilles, 1959, Le groupe de Québec et le groupe de Gaspé près du lac Weedon [Quebec]: Naturaliste Canadien, v. 86, no. 11, p. 243–263.

——1960a, Preliminary report on Weedon area, Wolfe and Compton electoral districts [Quebec]: Quebec Dept. Mines, Mineral Deposits Br., Prelim. Rept. 416, 9 p.

——1960b, Preliminary report on Gould area, Wolfe and Compton electoral districts [Quebec]: Quebec Dept. Mines, Mineral Deposits Br., Prelim. Rept. 432, 10 p.

Duquette, Gilles, 1961, Preliminary report on Lake Aylmer area, Wolfe and Frontenac Counties [Quebec]: Quebec Dept. Nat. Resources, Mineral Deposits Br., Prelim. Rept. 457, 13 p.

Eakins, P. R., 1964, Sutton map area, Quebec: Canada Geol. Survey Paper 63–34, 3 p.

Eaton, G. P., and Rosenfeld, J. L., 1960, Gravimetric and structural investigations in central Connecticut: 21st Internat. Geol. Cong., Copenhagen 1960, Rept., Part 2, 168–178.

Ells, R. W., 1887, Report on the geology of a portion of the Eastern Townships of Quebec, relating more especially to the counties of Compton, Stanstead, Beauce, Richmond, and Wolfe: Canada Geol. Survey 2d Ann. Rept., 1886, Part J, 70 p., and map [251, southeast quarter sheet (Sherbrooke sheet) of geol. map of the Eastern Townships, Province of Quebec].

——1888, Second report on the geology of a portion of the Province of Quebec: Canada Geol. Survey 3d Ann. Rept., 1887–88, pt. K, p. 1–114, and map [375 (pub. 1890), northeast quarter sheet (Quebec sheet) of geol. map of the Eastern Townships, Province of Quebec].

Emerson, B. K., 1917, Geology of Massachusetts and Rhode Island: U.S. Geol. Survey Bull. 597, 289 p.

Emmons, Ebenezer, 1888, Geology of the Montmorenci [reprinted from Am. Mag., v. 1, p. 146–150, 1841]: Am. Geologist, v. 2, no. 2, p. 94–100.

Engel, A. E. J., 1956, Apropos the Grenville, *in* Thompson, J. E., *Editor*, The Grenville problem: Royal Soc. Canada Spec. Pub. 1, p. 74–98.

——1963, Geologic evolution of North America: Science, v. 140, no. 3563, p. 143–152.

Engle, A. E. J., and Engel, C. G., 1953, General features of the Grenville Series, Part 1 *of* Grenville Series in the northwest Adirondack Mountains, New York: Geol. Soc. America Bull., v. 64, no. 9, p. 1013–1047.

Engel, A. E. J., Engel, C. G., and Havens, R. G., 1965, Chemical characteristics of oceanic basalts and the upper mantle: Geol. Soc. America Bull., v. 76, no. 7, p. 719–734.

Eric, J. H., and Dennis, J. G., 1958, Geology of the Concord-Waterford area, Vermont: Vermont Geol. Survey Bull. 11, 66 p.

Ern, E. H., Jr., 1963, Bedrock geology of the Randolph quadrangle, Vermont: Vermont Geol. Survey Bull. 21, 96 p.

Fairbairn, H. W., 1933, Chemical changes in metabasalt from southern Quebec: Jour. Geology, v. 41, no. 5, p. 553–558.

——1935, Notes on the mechanics of rock foliation: Jour. Geology, v. 43, no. 6, p. 591–608.

Fairbairn, H. W., Gaure, G., Pinson, W. H., Jr., Hurley, P. M., and Powell, J. L., 1963, Initial ratio of strontium 87 to strontium 86, whole-rock age, and discordant biotite in the Monteregian igneous province, Quebec: Jour. Geophys. Research, v. 68, no. 24, p. 6515–6522.

Fairbairn, H. W., Hurley, P. M., Pinson, W. H., Jr., and Cormier, R. F., 1960, Age of the granitic rocks of Nova Scotia: Geol. Soc. America Bull., v. 71, no. 4, p. 399–413.

Fairbairn, H. W., Moorbath, S., Ramo, A. O., Pinson, W. H., Jr., and Hurley, P. M., 1967, Rb-Sr age of granitic rocks of southeastern Massachusetts and the age of the Lower Cambrian at Hoppin Hill: Earth Planetary Sci. Letters., v. 2, no. 4, p. 321–328.

Faul, Henry, Stern, T. W., Thomas, H. H., and Elmore, P. L. D., 1963, Ages of intrusion and metamorphism in the northern Appalachians: Am. Jour. Sci., v. 261, no. 1, p. 1–19.

Fisher, D. W., 1968, Geology of the Plattsburg and Rouses Point, New York-Vermont quadrangles: Vermont Geol. Survey, Spec. Bull. 1, 51 p.

Fisher, D. W., and Hanson, G. F., 1951, Revisions in the geology of Saratoga Springs, New York and vicinity: Am. Jour. Sci., v. 249, no. 11, p. 795–814.

Fisher, D. W., Isachsen, Y. W., Rickard, L. V., Broughton, J. G., and Offield, T. W., 1961, Geologic map of New York: New York State Mus. and Sci. Service Geol. Survey Map and Chart Ser., no. 5.

Fitzpatrick, M. M., 1959, Gravity in the Eastern Townships of Quebec: Harvard Univ., Cambridge, Ph.D. dissert., 135 p.

Fleuty, M. J., 1964, Tectonic slides: Geol. Mag., v. 101, no. 5, p. 452–456.

Fowler, Phillip, 1950, Stratigraphy and structure of the Castleton area, Vermont: Vermont Geol. Survey Bull. 2, 83 p.

Freedman, Jacob, Wise, D. U., and Bentley, R. D., 1964, Pattern of folded folds in the Appalachian Piedmont along Susquehanna River: Geol. Soc. America Bull., v. 75, no. 7, p. 621–638.

Fritts, C. E., 1962, Age and sequence of metasedimentary and metavolcanic formations northwest of New Haven, Connecticut, *in* Geological Survey research 1962: U.S. Geol. Survey Prof. Paper 450-D, p. D32-D36.

——1963, Late Newark fault versus pre-Newark peneplain in Connecticut: Am. Jour. Sci., v. 261, no. 3, p. 268–281.

Gilluly, James, 1966, Orogeny and geochronology: Am. Jour. Sci., v. 264, no. 2, p. 97–111.

Glaessner, M. F., 1963a, The dating of the base of the Cambrian: Geol. Soc. India Jour., v. 4, p. 1–11.

——1963b, The base of the Cambrian: Geol. Soc. Australia Jour., v. 10, Part 1, p. 223–241.

Goldschmidt, V. M., 1922, Stammestypen der Eruptivgesteine: Vidensk. Skr. 1, Math.-naturw. Kl., 1922, no. 10, 12 p.

Goldsmith, Richard, 1964, Geologic map of New England; (1) general geology, (2) metamorphic zones, (3) radiometric ages: U.S. Geol. Survey open-file rept., 3 sheets, scale 1:1,000,000.

Goodwin, B. K., 1962, An alternate interpretation for the structure of east-central and northeastern Vermont: Pennsylvania Acad. Sci. Proc., v. 36, 1962, p. 200–207.

——1963, Geology of the Island Pond area, Vermont: Vermont Geol. Survey Bull. 20, 111 p.

Grant, J. A., 1964, Rubidium-strontium isochron study of the Grenville Front near Lake Timagami, Ontario: Science, v. 146, no. 3647, p. 1049–1053.

Grant, J. A., Wasserburg, G. J., and Albee, A. L., 1965, Rubidium-strontium isochron study of the Grenville Front near Lake Timagami, Ontario, Canada (Abs.), *in* Abstracts for 1964: Geol. Soc. America Spec. Paper 82, p. 76–77.

Green, J. C., 1964, Stratigraphy and structure of the Boundary Mountain anticlinorium in the Errol quadrangle, New Hampshire-Maine: Geol. Soc. America Spec. Paper 77, 78 p.

——1968, Geology of the Connecticut Lakes—Parmachenee area, New Hampshire and Maine: Geol. Soc. America Bull., v. 79, no. 11, p. 1601–1638.

Green, J. C., and Guidotti, C. V., 1968, The Boundary Mountains anticlinorium in northern New Hampshire and northwestern Maine, p. 255–266, *in* Zen, E-an, White, W. S., Hadley, J. J., and Thompson, J. B., Jr., *Editors*, Studies of Appalachian Geology: Northern and Maritime: New York, Interscience Publishers, 475 p.

Grenier, P. E., 1957, Beetz Lake area, electoral district of Saguenay [Quebec]: Quebec Dept. Mines, Geol. Survey Br., Geol. Rept. 73, 77 p.

Guidotti, C. V., 1965, Geology of the Bryant Pond quadrangle, Maine: Maine Geol. Survey Quad. Mapping Ser., no. 3, 116 p.

Hack, J. T., 1960, Interpretation of erosional topography in humid temperate regions: Am. Jour. Sci., v. 258-A (Bradley Volume), p. 80–97.

——1965, Geomorphology of the Shenandoah Valley, Virginia and West Virginia, and origin of the residual ore deposits: U.S. Geol. Survey Prof. Paper 484, 84 p.

Hadley, J. B., 1942, Stratigraphy, structure, and petrology of the Mt. Cube area, New Hampshire: Geol. Soc. America Bull., v. 53, no. 1, p. 113–176.

——1949, Mount Grace quadrangle, Massachusetts—bedrock geology: U.S. Geol. Survey Geol. Quad. Map GQ-3.

——1950, Geology of the Bradford-Thetford area, Orange County, Vermont: Vermont Geol. Survey Bull. 1, 36 p.

——1964, Correlation of isotopic ages, crustal heating, and sedimentation in the Appalachian region, *in* Tectonics of the southern Appalachians: Virginia Polytech. Inst. Dept. Geol. Sci. Mem. 1, p. 33–45.

Hall, L. M., 1959, The geology of the St. Johnsbury quadrangle, Vermont and New Hampshire: Vermont Geol. Survey Bull. 13, 105 p.

Hamilton, W. B., 1963, Metamorphism in the Riggins region, western Idaho: U.S. Geol. Survey Prof. Paper 436, 95 p.

Hamilton, W. B., and Myers, W. B., 1967, The nature of batholiths: U.S. Geol. Survey Prof. Paper 554-C, 30 p.

Harker, Alfred, 1939, Metamorphism, a study of the transformations of rock-masses, 2d ed.: London, Methuen and Co., Ltd., 362 p.

Harland, W. B., Smith, A. G., and Wilcock, B., 1964, *Editors*, The Phanerozoic time-scale, a symposium dedicated to Professor Arthur Holmes: *in* Geol. Soc. London Quart. Jour., v. 120s, 458 p.

Harper, C. T., 1968a, Isotopic ages from the Appalachians and their tectonic significance: Canadian Jour. Earth Sci. v. 5, no. 1, p. 50-59.

——1968b, Isotopic ages from the Appalachians and their tectonic significance: Reply to J. G. Dennis: Canadian Jour. Earth Sci. v. 5, no. 4, p. 961–962.

——1968c, Isotopic ages from the Appalachians and their tectonic significance: Reply to W. R. Church: Canadian Jour. Earth Sci. v. 5, no. 5, p. 1334–1335.

Harris, P. M., Farrar, E., MacIntyre, R. M., York, Derek, and Miller, J. A., 1965, Potassium-argon age measurements on two igneous rocks from the Ordovician system of Scotland: Nature, v. 205, no. 4969, p. 352–353.

Harwood, D. S., 1967, Radar imagery; Parmachenee Lake area, west-central Maine: NASA Earth Resources Survey Program Tech. Letter NASA-81, 7 p.

——1968, Fossiliferous pre-Silurian rocks in the Cupsuptic quadrangle, west-central Maine (Abs.): Geol. Soc. America Spec. Paper 115, p. 269.

Harwood, D. S., and Berry, W. B. N., 1967, Fossiliferous lower Paleozoic rocks in the Cupsuptic quadrangle, west-central Maine, *in* Geological Survey research 1967: U.S. Geol. Survey Prof. Paper 575-D, p. D16-D23.

Hast, N., 1967, The state of stresses in the upper part of the earth's crust: Eng. Geology, v. 2, no. 1, p. 5–17.

Hatch, N. L., Jr., 1963, The geology of the Dixville quadrangle, New Hampshire: New Hampshire Dept. Resources and Econ. Devel. Bull. 1, 81 p.

——1968, Isoclinal folding indicated by primary sedimentary structures in western Massachusetts, *in* Geological Survey Research, 1968: U.S. Geol. Survey Prof. Paper 600-D, p. D108-D114.

Hatch, N. L., Jr., Osberg, P. H., and Norton, S. A., 1967, Stratigraphy and structure of the east limb of the Berkshire anticlinorium, *in* New England Intercollegiate Geol. Conf. Guidebook 59th Ann. Mtg., Oct. 1967, Amherst, Mass., Field trips in the Connecticut Valley, Massachusetts: p. 7–16.

Hawkes, H. E., Jr., 1940, Structural geology of the Plymouth-Rochester area, Vermont: Massachusetts Inst. Technology, Cambridge, Mass., Ph.D. dissert., 186 p.

——1941, Roots of the Taconic fault in west-central Vermont: Geol. Soc. America Bull., v. 52, no. 5, p. 649–666.

Hawley, David, 1957, Ordovician shales and submarine slide breccias of northern Champlain Valley in Vermont: Geol. Soc. America Bull., v. 68, p. 55–194.

Heald, M. T., 1950, Structure and petrology of the Lovewell Mountain quadrangle, New Hampshire: Geol. Soc. America Bull., v. 61, no. 1, p. 43–89.

Heard, H. C., and Rubey, W. W., 1966, Tectonic implications of gypsum dehydration: Geol. Soc. America Bull., v. 77, no. 7, p. 741–760.

Hills, Allan, and Gast, P. W., 1964, Age of pyroxene-hornblende granitic gneiss of the eastern Adirondacks by the rubidium-strontium whole-rock method: Geol. Soc. America Bull., v. 75, no. 8, p. 759–766.

Hills, E. S., 1964, Elements of structural geology: New York, John Wiley & Sons, Inc., 483 p.

Hodgson, R. A., 1961a, Regional study of jointing in Comb Ridge-Navajo Mountain area, Arizona and Utah: Am. Assoc. Petroleum Geologists Bull., v. 45, no. 1, p. 1–38.

——1961b, Classification of structures on joint surfaces: Am. Jour. Sci., v. 259, no. 7, p. 493–502.

Hofmann, H. J., 1963, Ordovician Chazy Group in southern Quebec: Am. Assoc. Petroleum Geologists Bull., v. 47, no. 2, p. 270–301.

Hollingworth, S. E., 1962, Depth and tectonics as factors in regional metamorphism (Symp.): Geol. Soc. London Proc., no. 1594, 1961–1962, p. 13–36.

Holm, J. L., and Kleppa, O. J., 1966, The thermodynamic properties of the aluminum silicates: Am. Mineralogist, v. 51, nos. 11–12, p. 1608–1622.

Holmes, Arthur, 1947, The construction of a geological time-scale: Geol. Soc. Glasgow Trans., v. 21, Part 1, 1945–1946, p. 117–152.

——1959, A revised geological time scale: Edinburgh Geol. Soc. Trans., v. 17, Part 3, p. 183–216.

Hubbert, M. K., 1928, The direction of the stresses producing given geologic strains: Jour. Geology, v. 36, no. 1, p. 75–84.

——1937, Theory of scale models as applied to the study of geologic structures: Geol. Soc. America Bull., v. 48, no. 10, p. 1459–1519.

Hudson, G. H., and Cushing, H. P., 1931, The dikes of Valcour Island and of the Peru and Plattsburg coast line (New York): New York State Mus. Bull. 286, p. 100–109.

Hurley, P. M., Boucot, A. J., Albee, A. L., Paul, Henry, Pinson, W. H., Jr., and Fairbairn, H. W., 1959, Minimum age of the Lower Devonian slate near Jackman, Maine: Geol. Soc. America Bull., v. 70, no. 7, p. 947–949.

Hurley, P. M., Fairbairn, H. W., Pinson, W. H., Jr., and Faure, G., 1960a, K-A and Rb-Sr minimum ages for the Pennsylvanian section in the Narragansett Basin [Massachusetts and Rhode Island]: Geochim. et Cosmochim. Acta, v. 18, nos. 3–4, p. 247–258.

Hurley, P. M., Fairbairn, H. W., Pinson, W. H., Jr., and Hower, J., 1960b, Redistribution of radiogenic Sr^{87} between Rb-rich and Rb-poor phases during metamorphism, *in* Variations in isotopic abundances of strontium, calcium, and argon and related topics: U.S. Atomic Energy Comm. [Pubs.] NYO-3941, 8th Ann. Prog. Rept. 1960, Contract AT (30–1)-1381, p. 225–235.

Hurley, P. M., Fairbairn, H. W., Pinson, W. H., Jr., Hower, J., Hughes, H., 1961a, Progress report on Rb-Sr petrology and geochronology of the Monteregian Hills, Quebec, *in* Variations in isotopic abundances of strontium, calcium, and argon and related topics: U.S. Atomic Energy Comm. [Pubs.] NYO-3942, 9th Ann. Prog. Rept. 1961, Contract AT (30–1)-1381, p. 151–160.

——1961b, Observed migration of Sr^{87} in metamorphic rocks in Vermont, *in* Variations in isotopic abundances of strontium, calcium, and argon and related topics: U.S. Atomic Energy Comm. [Pubs.] NYO-3942, 9th Ann. Prog. Rept. 1961, Contract AT (30–1)-1381, p. 187–191.

Innes, M. J. S., and Argun-Weston, A., 1967, Gravity measurements in Appalachia and their structural implications: Royal Soc. Canada Spec. Pub. 10, p. 69–83.

Isachsen, Y. W., 1964, Extent and configuration of the Precambrian in northeastern United States: New York Acad. Sci. Trans., Ser. 2, v. 26, no. 7, p. 812–829.

Jahns, R. H., 1943, Sheet structure in granites, its origin and use as a measure of glacial erosion in New England: Jour Geology, v. 51, no. 2, p. 71–98.

——1967, Serpentinites of the Roxbury district, Vermont, p. 135–160, Part 2 *of* Chap. 5, Alpine-type ultramafic associations, *in* Wyllie, P. J., *Editor*, Ultramafic and Related Rocks: New York, John Wiley & Sons, Inc., 464 p.

James, H. L., 1958, Stratigraphy of pre-Keweenawan rocks in parts of northern Michigan: U.S. Geol. Survey Prof. Paper 314-C, p. 27–44.

Johansson, W. I., 1963, Geology of the Lunenburg-Brunswick-Guildhall area, Vermont: Vermont Geol. Survey Bull. 22, 86 p.

Johnson, D. W., 1931, Stream sculpture on the Atlantic slope, a study in the evolution of Appalachian rivers: New York, Columbia Univ. Press, 142 p.

Joyner, W. B., 1963, Gravity in north-central New England: Geol. Soc. America Bull., v. 74, no. 7, p. 831–857.

Kay, Marshall, 1951, North American geosynclines: Geol. Soc. America Mem. 48, 143 p.

Keith, Arthur, 1923a, Cambrian succession of northwestern Vermont: Am. Jour. Sci., 5th ser., v. 5, p. 97–139.

——1923b, Outlines of Appalachian structure: Geol. Soc. America Bull., v. 34, no. 2, p. 309–380.

——1932, Stratigraphy and structure of northwestern Vermont: Washington Acad. Sci. Jour., v. 22, no. 13, p. 357–379, 393–406.

——1933, Outline of the structure and stratigraphy of northwestern Vermont: 16th Internat. Geol. Cong., United States, Guidebook 1, p. 48–61.

King, P. B., 1950, Geology of the Elkton area, Virginia: U.S. Geol. Survey Prof. Paper 230, 82 p.

Knox, J. K., 1917, Southwestern part of the Thetford-Black Lake mining district (Coleraine sheet) [Quebec]: Canada Geol. Survey Summ. Rept. 1916, p. 229–245.

Konig, R. H., 1961, Geology of the Plainfield quadrangle, Vermont: Vermont Geol. Survey Bull. 16, 86 p.

Konig, R. H., and Dennis, J. G., 1964, The geology of the Hardwick area, Vermont: Vermont Geol. Survey Bull. 24, 57 p.

Kranck, E. H., 1961, The tectonic position of the anorthosites of eastern Canada: Soc. Géol. Finlande Comptes rendus, no. 33, p. 299–320.

Kulp, J. L., 1961, Geologic time scale: Science, v. 133, no. 3459, p. 1105–1114.

Kumarapeli, P. S., and Saull, V. A., 1966, The St. Lawrence valley system; a North American equivalent of the East African rift valley system: Canadian Jour. Earth Sci., v. 3, no. 5, p. 639–658.

Kuno, Hisashi, 1960, High-alumina basalt: Jour. Petrology, v. 1, no. 2, p. 121–145.

Lajoie, Jean, Lespérance, P. J., and Béland, J. R., 1968, Silurian stratigraphy and paleogeography of Matapedia-Témiscouata region, Quebec: Am. Assoc. Petroleum Geologists Bull., v. 52, no. 4, p. 615–640.

Lamarche, R. Y., 1962, Etude des conglomérats de la région d'Orford-Sherbrooke [Quebec]: Laval Univ., Quebec, M.S. thesis, 70 p.

——1965, Géologie de la Région de Sherbrooke, Comté de Sherbrooke, Quebec: Laval Univ., Quebec, D.Sc. thesis, 291 p.

——1967, Geology of Beauvoir-Ascot Corner area, Sherbrooke, Richmond, and Compton Counties [Quebec]: Quebec Dept. Natl. Resources, Mineral Deposits Service, Prelim. Rept. 560, 16 p.

Lapham, D. M., and Bassett, W. A., 1964, K-Ar dating of rocks and tectonic events in the Piedmont of southeastern Pennsylvania: Geol. Soc. America Bull., v. 75, no. 7, p. 661–667.

Lapham, D. M., and McKague, H. L., 1964, Structural patterns associated with the serpentinites of southeastern Pennsylvania: Geol. Soc. America Bull., v. 75, no. 7, p. 639–659.

Larochelle, Andre, 1968, Palaeomagnetism of the Monteregian Hills: New results: Jour. Geophys. Research, v. 73, no. 10, p. 3239–3246.

Laverdière, J. W., 1936, Marbleton and vicinity, Dudswell Township, Wolfe County [Quebec]: Quebec Bur. Mines Ann. Rept. 1935, Part D, p. 29–40.

Lespérance, P. J., 1959, Preliminary report on Squateck area (west half), Témiscouata, Rivière-du-Loup and Rimouski electoral districts [Quebec]: Quebec Dept. Mines, Geol. Survey Br., Prelim. Rept. 385, 10 p.

——1963, Preliminary report on Acton area, Bagot and Shefford Counties [Quebec]: Quebec Dept. Nat. Resources, Geol. Survey Br., Prelim. Rept. 496, 8 p.

Lespérance, P. J., and Greiner, H. R., (in press), Squateck-Cabano area, Rimouski, Rivière-du-Loup and Témiscouata Counties [Quebec]: Quebec Dept. Nat. Resources Geol. Rept.

Logan, W. E., 1862, Considerations relating to the Quebec Group, and the upper copper-bearing rocks of Lake Superior: Am. Jour. Sci., 2d ser., v. 33, p. 320–327.

Logan, W. E., and others, 1863, Geology of Canada: Canada Geol. Survey Prog. Rept. to 1863, 983 p.

Long, L. E., 1962, Isotopic age study, Dutchess County, New York: Geol. Soc. America Bull., v. 73, no. 8, p. 997–1005.

Long, L. E., and Kulp, J. L., 1962, Isotopic age study of the metamorphic history of the Manhattan and Reading Prongs: Geol. Soc. America Bull., v. 73, no. 8, p. 969–995.

Lovering, J. F., and Richards, J. R., 1964, Potassium-argon age study of possible lower-crust and upper-mantle inclusions in deep-seated intrusions: Jour. Geophys. Research, v. 69, no. 22, p. 4895–4901.

Lowdon, J. A., *Compiler*, 1960, Isotopic ages, Rept. 1 *in* Age determinations by the Geological Survey of Canada: Canada Geol. Survey Paper 60–17, 51 p.

——1961, Isotopic ages, Rept. 2 *in* Age determinations of the Geological Survey of Canada: Canada Geol. Survey Paper 61–17, 127 p.

——1963a, Isotopic ages, Rept. 3 *in* Age determinations by the Geological Survey of Canada: Canada Geol. Survey Paper 62–17, p. 5–122.

Lowdon, J. A., 1963b, Isotopic ages, Rept. 4 *in* Age determinations by the Geological Survey of Canada: Canada Geol. Survey Paper 63-17, p. 5–124.

Lundgren Lawrence, Jr., 1962, Deep River area, Connecticut—stratigraphy and structure: Am. Jour. Sci., v. 260, no. 1, p. 1–23.

Lyons, J. B., 1955, Geology of the Hanover quadrangle, New Hampshire-Vermont: Geol. Soc. America Bull., v. 66, no. 1, p. 105–145.

Lyons, J. B., and Faul, Henry, 1968, Isotope geochronology of the northern Appalachians, p. 305–318, *in* Zen, E-an, White, W. S., Hadley, J. J., and Thompson, J. B., Jr., *Editors,* Studies of Appalachian Geology: Northern and Maritime: New York, Interscience Publishers, 475 p.

Lyons, J. B., Jaffe, H. W., Gottfried, David, and Warning, C. L., 1957, Lead-alpha ages of some New Hampshire granites: Am. Jour. Sci., v. 255, no. 8, p. 527–546.

MacClintock, Paul, and Stewart, D. P., 1965, Pleistocene geology of the St. Lawrence lowland: New York State Mus. and Sci. Service Bull. 394, 152 p.

MacDonald, G. A., and Katsura, Takashi, 1964, Chemical composition of Hawaiian lavas: Jour. Petrology, v. 5, no. 1, p. 82–133.

MacFadyen, J. A., Jr., 1956, The geology of the Bennington area, Vermont: Vermont Geol. Survey Bull. 7, 72 p.

MacIntyre, R. M., Derek, York, and Moorhouse, W. W., 1967, Potassium-argon age determinations in the Madoc-Bancroft area in the Grenville province of the Canadian Shield: Canadian Jour. Earth Sci., v. 4, no. 5, p. 815–828.

Marleau, R. A., 1958a, Geology of the Woburn, East Megantic, and Armstrong areas, Frontenac and Beauce Counties (Quebec]: Laval Univ., Quebec, D.Sc. thesis, 184 p.

——1958b, Preliminary report on East Megantic and Armstrong areas, electoral districts of Frontenac and Beauce (Quebec): Quebec Dept. Mines, Geol. Survey Br., Prelim. Rept. 362, 7 p.

——1959, Age relations in the Lake Megantic Range, southern Quebec: Geol. Assoc. Canada Proc., v. 11, p. 129–139.

——1968, Woburn–East Megantic–Armstrong area, Frontenac and Beauce Counties [Quebec]: Quebec Dept. Nat. Resources, Geol. Rept. 131, 55 p.

Maxwell, J. C., 1962, Origin of slaty and fracture cleavage in the Delaware Water Gap area, New Jersey and Pennsylvania, *in* Petrologic Studies—A Volume in Honor of A. F. Buddington: Geol. Soc. America, p. 281–311.

McDonald, B. C., 1968, Deglaciation and differential postglacial rebound in the Appalachian region of southeastern Quebec: Jour. Geol., v. 76, no. 6, p. 664–677.

McGerrigle, H. W., 1930, Three geological series in northwestern Vermont: Vermont State Geologist 17th Bienn. Rept. 1929–1930, p. 179–191.

——1934, Western Témiscouata, with parts of Kamouraska and Rivière-du-Loup Counties [Quebec]: Quebec Bur. Mines Ann. Rept. 1933, Part D, p. 93–129.

——1954, The Tourelle and Courcellette areas, Gaspé Peninsula [Quebec]: Quebec Dept. Mines, Geol. Survey Br., Geol. Rept. 62, 63 p.

Melihercsik, S. J., 1954, A history of the formation names in the Quebec group with special reference to the Charny formation: Naturaliste Canadien, v. 81, nos. 8–9, p. 165–180.

Miyashiro, Akiho, 1961, Evolution of metamorphic belts: Jour. Petrology, v. 2, no. 3, p. 277–311.

——1967, Orogeny, regional metamorphism, and magmatism in the Japanese Islands: Dansk Geol. Foren. Medd., v. 17, no. 4, p. 390–446.

Moench, R. H., 1966, Relation of S_2 schistosity to metamorphosed clastic dikes, Rangeley-Phillips area, Maine: Geol. Soc. America Bull., v. 77, no. 12, p. 1449–1461.

Moorbath, S., 1967, Recent advances in the application and interpretation of radiometric age data: Earth-Sci. Reviews, v. 3, no. 3, p. 111–133.

Moore, G. E., Jr., 1949, Structure and metamorphism of the Keene-Brattleboro area, New Hampshire-Vermont: Geol. Soc. America Bull., v. 60, no. 10, p. 1613–1669.

Murthy, V. R., 1957, Bed rock geology of the East Barre area, Vermont: Vermont Geol. Survey Bull. 10, 121 p.

——1958, A revision of the lower Paleozoic stratigraphy in eastern Vermont: Jour. Geology, v. 66, no. 3, p. 276–287.

——1959, A revision of the lower Paleozoic stratigraphy in eastern Vermont; a reply to discussion by Walter S. White: Jour. Geology, v. 67, no. 5, p. 581–582.

Myers, P. B., Jr., 1964, Geology of the Vermont portion of the Averill quadrangle, Vermont: Vermont Geol. Survey Bull. 27, 69 p.

Naylor, R. S., 1967, A field and geochronologic study of mantled gneiss domes in central New England: California Inst. Technology, Pasadena, Calif., Ph.D. dissert., 123 p.

——1968, Origin and regional relationships of the core-rocks of the Oliverian domes, p. 231–240, *in* Zen, E-an, White, W. S., Hadley, J. J., and Thompson, J. B., Jr., *Editors*, Studies of Appalachian Geology: Northern and Maritime: New York, Interscience Publishers, 475 p.

——1969, Age and origin of the Oliverian Domes, central western New Hampshire: Geol. Soc. America Bull., v. 80, p. 405–427.

Naylor, R. S., and Boucot, A. J., 1965, Origin and distribution of rocks of Ludlow age (Late Silurian) in the northern Appalachians: Am. Jour. Sci., v. 263, no. 2, p. 153–169.

Naylor, R. S., and Wasserburg, G. J., 1966, Rb-Sr studies on mantled gneiss domes in central New England (Abs.): Am. Geophys. Union Trans., v. 47, no. 1, p. 194–195.

Neale, E. R. W., Béland, J. R., Potter, R. R., and Poole, W. H., 1961, A preliminary tectonic map of the Canadian Appalachian region based on age of folding: Canadian Mining and Metall. Bull., v. 54, no. 593, p. 687–694.

Neuman, R. B., and Rankin, D. W., 1966, Bedrock geology of the Shin Pond region [Maine], *in* New England Intercollegiate Geol. Conf. Guidebook 58th Ann. Mtg., Mount Katahdin, Maine, Oct., 1966: p. 8–16.

Newton, R. C., 1966, Kyanite-sillimanite equilibrium at 750°C: Science, v. 151, no. 3715, p. 1222–1225.

Noble, D. C., and Billings, M. P., 1967, Pyroclastic rocks of the White Mountain Magma Series, New Hampshire: Nature, v. 216, no. 5118, p. 906–907.

Ollerenshaw, N. C., 1968, Cuoq–Langis area, Matane and Matapédia Counties [Quebec]: Quebec Dept. Nat. Resources, Geol. Rept. 121, 192 p.

Olsen, E. J., 1961, High temperature acid rocks associated with serpentinite in eastern Quebec: Am. Jour. Sci., v. 259, no. 5, p. 329–347.

Opdyke, N. D., and Wensink, H., 1966, Paleomagnetism of rocks from the White Mountain plutonic-volcanic series in New Hampshire and Vermont: Jour. Geophys. Research, v. 71, no. 12, p. 3045–3051.

Osberg, P. H., 1952, The Green Mountain anticlinorium in the vicinity of Rochester and East Middlebury, Vermont: Vermont Geol. Survey Bull. 5, 127 p.

——1956, Stratigraphy of the Sutton Mountains, Quebec; key to stratigraphic correlation in Vermont (Abs.): Geol. Soc. America Bull., v. 67, no. 12, pt. 2, p. 1820.

Osberg, P. H., 1959, The stratigraphy and structure of the Coxe Mountain area, Vermont, *in* New England Intercollegiate Geol. Conf. Guidebook 51st Ann. Mtg., Rutland, Vermont., Oct. 1959, Stratigraphy and structure of west-central Vermont and adjacent New York: p. 45–52.

——1965, Structural geology of the Knowlton-Richmond area, Quebec: Geol. Soc. America Bull., v. 76, no. 2. p. 223–250.

——(in press), Lower Paleozoic stratigraphy and structural geology of the Green Mountain-Sutton Mountain anticlinorium, Vermont and southern Quebec, *in* Kay, Marshall, *Editor*, Symposium on the North Atlantic—geology and continental drift, Gander meeting, Am. Assoc. Petroleum Geologists Mem. 12.

Osberg, P. H., Moench, R. H., and Warner, Jeffrey, 1968, Stratigraphy of the Merrimack synclinorium in west-central Maine, p. 241–253, *in* Zen, E-an, White, W. S., Hadley, J. J., and Thompson, J. B., Jr., *Editors*, Studies of Appalachian Geology: Northern and Maritime: New York, Interscience Publishers, 475 p.

Osborne, F. F., 1937, Magma and ore deposits: Royal Soc. Canada Trans., 3d ser., v. 31, Sec. 4, p. 121–128.

——1956, Geology near Quebec city: Naturaliste Canadien, v. 83, nos. 8–9, p. 157–223.

Osborne, F. F., and Berry, W. B. N., 1966, Tremadoc rocks at Lévis and Lauzon: Naturaliste Canadien, v. 93, no. 2, p. 133–143.

Osborne, F. F., and Morin, Marcel, 1962, Tectonics of part of the Grenville subprovince in Quebec, *in* The tectonics of the Canadian Shield: Royal Soc. Canada Spec. Pub. 4, p. 118–143.

Osborne, F. F., and Riva, John, 1966, Post-Lévis beds of the Quebec Group at St. Apollinaire, Lotbinière County, Province of Quebec: Naturaliste Canadien, v. 93, no. 2, p. 145–151.

Page, L. R., 1937, The geology of the Rumney quadrangle, New Hampshire: Univ. Minnesota Ph.D. dissert., 150 p.

——1968, Devonian plutonic rocks in New England, p. 371–383, *in* Zen, E-an, White, W. S., Hadley, J. J., and Thompson, J. B., Jr., *Editors*, Studies of Appalachian Geology: Northern and Maritime: New York, Interscience Publishers, 475 p.

Parker, J. M., III, 1942, Regional systematic jointing in slightly deformed sedimentary rocks: Geol Soc. America Bull., v. 53, no. 3, p. 381–408.

Pavlides, Louis, and Berry, W. B. N., 1966, Graptolite-bearing Silurian rocks of the Houlton-Smyrna Mills area, Aroostook County, Maine, *in* Geological Survey research 1966: U.S. Geol. Survey Prof. Paper 550-B, p. B51-B61.

Pavlides, Louis, Boucot, A. J., and Skidmore, W. B., 1968, Stratigraphic evidence for the Taconic orogeny in the northern Appalachians, p. 61–82, *in* Zen, E-an, White, W. S., Hadley, J. J., and Thompson, J. B., Jr., *Editors*, Studies of Appalachian Geology: Northern and Maritime: New York, Interscience Publishers, 475 p.

Pavlides, Louis, Menchner, Ely, Naylor, R. S., and Boucot, A. J., 1964, Outline of the stratigraphy and tectonic features of northeastern Maine, *in* Geological Survey research 1964: U.S. Geol. Survey Prof. Paper 501-C, p. C28-C38.

Phillips, F. C., 1937, A fabric study of some Moine schists and associated rocks: Geol. Soc. London Quart. Jour., v. 93, Part 4, p. 581–620.

Philpotts, A. R., and Miller, J. A., 1963, A Pre-Cambrian glass from St. Alexis-des-Monts, Quebec: Geol. Mag., v. 100, no. 4, p. 337–344.

Pierce, K. L., 1965, Geomorphic significance of a Cretaceous deposit in the Great Valley of southern Pennsylvania, *in* Geological Survey research 1965: U.S. Geol. Survey Prof. Paper 525-C, p. C152-C156.

Platt, L. B., 1962, Fluid pressure in thrust faulting, a corollary: Am. Jour. Sci., v. 260, no. 2, p. 107-114.

Poole, W. H., 1967, Tectonic evolution of Appalachian region of Canada: Geol. Assoc. Canada Spec. Paper 4, p. 9–51.

Poole, W. H., Béland, J. R., and Wanless, R. K., 1963, Minimum age of Middle Ordovician rocks in southern Quebec: Geol. Soc. America Bull., v. 74, no. 8, p. 1063–1065.

——1964, Minimum age of Middle Ordovician rocks in southern Quebec—Reply: Geol. Soc. America Bull., v. 75, no. 9, p. 911.

Poole, W. H., Kelley, D. G., and Neale, E. R. W., 1964, Age and correlation problems in the Appalachian region of Canada: Royal Soc. Canada Spec. Pub. 8, p. 61–84.

Potter, D. B., 1963, Stratigraphy and structure of the Hoosick Falls area, *in* Geol. Soc. America Guidebook Field Trip 3, Stratigraphy structure, sedimentation, and paleontology of the southern Taconic region, eastern New York: p. 58–67.

——1968, Giant submarine slide blocks beneath east edge of Taconic allochthon, New York (Abs.): Geol. Soc. America Spec. Paper 115, p. 285.

Price, N. J., 1959, Mechanics of jointing in rocks: Geol. Mag., v. 96, no. 2, p. 149–167.

Quebec Department of Mines, 1957, Geologic map, Province of Quebec: Quebec Dept. Mines Map 1129.

Quinn, A. W., Jaffe, H. W., Smith, W. L., and Waring, C. L., 1957, Lead-alpha ages of Rhode Island granitic rocks compared to their geologic ages: Am. Jour. Sci., v. 255, no. 8, p. 547–560.

Quinn, A. W., and Moore, G. E., Jr., 1968, Sedimentation, tectonism, and plutonism of the Narragansett Basin region, p. 269–279, *in* Zen, E-an, White, W. S., Hadley, J. J., and Thompson, J. B., Jr., *Editors*, Studies of Appalachian Geology: Northern and Maritime: New York, Interscience Publishers, 475 p.

Quinn, A. W., and Oliver, W. A., Jr., 1962, Pennsylvanian rocks of New England, *in* Branson, C. C., *Editor*, Pennsylvanian System in the United States—A symposium: Tulsa, Oklahoma, Am. Assoc. Petroleum Geologists, p. 60–73.

Raleigh, C. B., 1967, Experimental deformation of ultramafic rocks and minerals, p. 191–199, Part 4 of Chap. 6, Deformation of alpine ultramafic rocks, *in* Wyllie, P. J., *Editor*, Ultramafic and related rocks: New York, John Wiley & Sons, Inc., 464 p.

Raleigh, C. B., and Paterson, M. S., 1965, Experimental deformation of serpentinite and its tectonic implications: Jour. Geophys. Research, v. 70, no. 16, p. 3965–3985.

Rankin, D. W., 1968, Volcanism related to tectonism in the Piscataquis volcanic belt, an island arc of Early Devonian age in north-central Maine, p. 355–369, *in* Zen, E-an, White, W. S., Hadley, J. J., and Thompson, J. B., Jr., *Editors*, Studies of Appalachian Geology: Northern and Maritime: New York, Interscience Publishers, 475 p.

Rast, N., 1964, Morphology and interpretation of folds—a critical essay: Geol. Jour., v. 4, Part 1, p. 177–188.

Rickard, M. J., 1961, A note on cleavage in crenulated rocks: Geol. Mag., v. 98, no. 4, p. 324–332.

——1964a, Metamorphic tourmaline overgrowths in the Oak Hill Series of southern Quebec: Canadian Mineralogist, v. 8, Part 1, p. 86–91.

——1964b, Minimum age of Middle Ordovician rocks in southern Quebec: Discussion of reply: Geol. Soc. America Bull., v. 75, no. 9, p. 913–914.

——1965, Taconic orogeny in the western Appalachians—experimental application of microtextural studies to isotopic dating: Geol. Soc. America Bull., v. 76, no. 5, p. 523–535.

Riecker, R. E., and Rooney, T. P., 1966, Weakening of dunite by serpentine dehydration: Science, v. 152, no. 3719, p. 196–198.

Riordon, P. H., 1954, Preliminary report on Thetford Mines-Black Lake area, Frontenac, Megantic, and Wolfe Counties (Quebec): Quebec Dept. Mines, Mineral Deposits Br., Prelim. Rept. 295, 23 p.

——1957, Evidence of a pre-Taconic orogeny in southeastern Quebec: Geol. Soc. America Bull., v. 68, no. 3, p. 389–394.

——1962, Geology of the asbestos belt in southeastern Quebec: Canada Mining and Metall. Bull., v. 55, no. 601, p. 311–313, 500.

Riva, John, 1966a, Upper Lévis graptolites from Cowansville, southern Quebec: Jour. Paleontology, v. 40, no. 1, p. 220–221.

——1966b, New assemblage of Middle Ordovician graptolites from the Appalachian region, Quebec: Naturaliste Canadien, v. 93, no. 2, p. 153–156.

Roberts, J. C., 1961, Feather-fracture, and the mechanics of rock-jointing: Am. Jour. Sci., v. 259, no. 7, p. 481–492.

Robinson, Peter, 1966, Alumino-silicate polymorphs and Paleozoic erosion rates in central Massachusetts (Abs.): Am. Geophys. Union Trans., v. 47, no. 2, p. 424.

——1967a, Progress of bedrock geologic mapping in west-central Massachusetts, p. 29–44, *in* Farquhar, O. C., *Editor*, Economic Geology in Massachusetts, Proc. Conf. Jan. 1966: Amherst Mass., Univ. Massachusetts Grad. School, 568 p.

——1967b, Gneiss domes and recumbent folds of the Orange area, west central Massachusetts and New Hampshire, *in* New England Intercollegiate Geol. Conf. Guidebook 59th Ann. Mtg., Amherst, Mass., Oct. 1967, Field trips in the Connecticut Valley, Massachusetts: p. 17–47.

Robison, R. A., 1964, Middle-Upper Cambrian boundary in North America: Geol. Soc. America Bull., v. 75, no. 10, p. 987–993.

Rodgers, John, 1968, The eastern edge of the North American continent during the Cambrian and Early Ordovician, p. 141–149, *in* Zen, E-an, White, W. S., Hadley, J. J., and Thompson, J. B., Jr., *Editors*, Studies of Appalachian Geology: Northern and Maritime: New York, Interscience Publishers, 475 p.

Rodgers, John, Gates, R. M., Cameron, E. N., and Ross, R. J., Jr., 1956, Preliminary geological map of Connecticut: Connecticut Geol. and Nat. History Survey.

Rodgers, John, Gates, R. M., and Rosenfeld, J. L., 1959, Explanatory text for preliminary geological map of Connecticut, 1956: Connecticut Geol. and Nat. History Survey Bull. 84, 64 p.

Rooney, T. P., and Riecker, R. E., 1966, High-pressure reactions and shear strength of serpentinized dunite, reply to Sclar and Carrison: Science, v. 153, no. 3741, p. 1287.

Rose, H. S., Jr., and Stern, T. W., 1960, Spectrochemical determination of lead in zircon for lead-alpha age measurements: Am. Mineralogist, v. 45, nos. 11–12 p. 1243–1256.

Rosenfeld, J. L., 1960, Rotated garnets and the diastrophic-metamorphic sequence in southeastern Vermont (Abs.): Geol. Soc. America Bull., v. 71, no. 12, Part 2, p. 1960.

——1965, Further evidence of the nature of the major diastrophism in southeast Vermont (Abs.), *in* Abstracts for 1964: Geol. Soc. America Spec. Paper 82, p. 167.

——1968, Garnet rotations due to the major Paleozoic deformations in southeast Vermont, p. 185–202, *in* Zen, E-an, White, W. S., Hadley, J. J., and Thompson, J. B., Jr., *Editors*, Studies of Appalachian Geology: Northern and Maritime: New York, Interscience Publishers, 475 p.

Roth, Horst, 1965, A structural study of the Sutton Mountains, Quebec: McGill Univ., Montreal, Ph.D. dissert., 139 p.

Rubey, W. W., and Hubbert, M. K., 1959, Overthrust belt in geosynclinal area of western Wyoming in light of fluid-pressure hypothesis, Part 2 *of* Role of fluid pressure in mechanics of overthrust faulting: Geol. Soc. America Bull., v. 70, no. 2, p. 167–205.

——1965, Role of fluid pressure in mechanics of overthrust faulting—Reply: Geol. Soc. America Bull., v. 76, no. 4, p. 469–474.

Rutland, R. W. R., 1966, Discussion, *in* Chinner, G. A., The distribution of pressure and temperature during Dalradian metamorphism: Geol. Soc. London Quart. Jour., v. 122, Part 2, p. 185–186.

St. Julien, Pierre, 1961a, Preliminary report on Fraser Lake area, Shefford and Stanstead Counties [Quebec]: Quebec Dept. Mines, Mineral Deposits Br., Prelim. Rept. 439. 11 p.

——1961b, Preliminary report on Lac Montjoie area, Sherbrooke, Richmond, and Stanstead Counties [Quebec]: Quebec Dept. Nat. Resources, Mineral Deposits Br., Prelim. Rept. 464, 14 p.

——1963a, Géologie de la Région Orford-Sherbrooke [Quebec]: Laval Univ. D.Sc. thesis, 369 p.

——1963b, Preliminary report on Saint Elie d'Orford area, Sherbrooke and Richmond Counties [Quebec]: Quebec Dept. Nat. Resources, Mineral Deposits Br., Prelim. Rept. 492, 14 p.

——1965, (Geological map) Orford-Sherbrooke area, Richmond, Sherbrooke, Shefford, Brome, Stanstead, and Compton Counties [Quebec]: Quebec Dept. Nat. Resources, Mineral Deposits Services, Map 1619.

——1967, Tectonics of part of the Appalachian region of southeastern Quebec (southwest of the Chandière River): Royal Soc. Canada Spec. Pub. 10, p. 41–47.

St. Julien, Pierre, and Lamarche, R. Y., 1965, Geology of Sherbrooke area, Sherbrooke County [Quebec]: Quebec Dept. Nat. Resources, Mineral Deposits Service, Prelim. Rept. 530, 34 p.

Sanders, J. E., 1963, Late Triassic tectonic history of northeastern United States: Am. Jour. Sci., v. 261, no. 6, p. 501–524.

Schumm, S. A., 1963, The disparity between the present rates of denudation and orogeny: U.S. Geol. Survey Prof. Paper 454-H, p. H1-H13.

Sclar, C. B., and Carrison, L. C., 1966, High-pressure reactions and shear strength of serpentinized dunite: Science, v. 153, no. 3741, p. 1285–1286.

Secor, D. T., Jr., 1965, Role of fluid pressure in jointing: Am. Jour. Sci., v. 263, no. 8, p. 633–646.

Shackleton, R. M., 1958, Downward-facing structures of the Highland Border: Geol. Soc. London Quart. Jour., v. 113, Part 3, p. 361–392.

Sharp, H. S., 1929, The physical history of the Connecticut shore line: Connecticut Geol. and Nat. History Survey Bull. 46, 97 p.

Shaw, A. B., 1958, Stratigraphy and structure of the St. Albans area, northwestern Vermont: Geol. Soc. America Bull., v. 69, no. 5, p. 519–567.

——1961, Cambrian of southeastern and northwestern New England, *in* Problemas generales, Europa occidental, África, URSS, Asia, América (symp., Part 3): Internat. Geol. Cong., 20th, Mexico City 1956 [Moscow, D. I. Shcherbakov] p. 433–471.

——1966a, Paleontology of northwestern Vermont. X. Fossils from the (Cambrian) Skeels Corners Formation: Jour. Paleontology, v. 40, no. 2, p. 269–295.

——1966b, Paleontology of northwestern Vermont. XI. Fossils from the Middle Cambrian St. Albans Shale: Jour. Paleontology, v. 40, no. 4, p. 843–858.

Shimazu, Yasuo, 1961, A geophysical study of regional metamorphism: Japanese Jour. Geophysics, v. 2, no. 4, p. 135–176.

Silver, L. T., 1964, Isotope investigations of zircons in Precambrian igneous rocks of the Adirondack Mountains, New York (Abs.): Geol. Soc. America Spec. Paper 76, p. 150–151.

——1965, U-Pb isotopic data in zircons of the Grenville Series of the Adirondack Mountains, New York (Abs.): Am. Geophys. Union Trans., v. 46, no. 1, p. 164.

Skehan, J. W., 1961, The Green Mountain anticlinorium in the vicinity of Wilmington and Woodford, Vermont: Vermont Geol. Survey Bull. 17, 159 p.

Smith, D. G. W., Baadsgaard, Halfdan, Folinsbee, R. E., and Lipson, J. I., 1961, K/Ar age of Lower Devonian bentonites of Gaspé, Quebec, Canada: Geol. Soc. America Bull., v. 72, no. 1, p. 171–173.

Snyder, G. L., 1964, Petrochemistry and bedrock geology of the Fitchville quadrangle, Connecticut: U.S. Geol. Survey Bull. 1161-I, 63 p.

Stanley, R. S., 1964, The bedrock geology of the Collinsville quadrangle (Connecticut): Connecticut Geol. and Nat. History Survey Quad. Rept. 16, 99 p.

——1967, Geometry and relations of some minor folds and their relation to the Woronoco nappe, Blandford and Woronoco quadrangles, Massachusetts, *in* New England Intercollegiate Geol. Conf. Guidebook, 59th Ann. Mtg., Amherst, Mass., Oct. 1967, Field trips in the Connecticut Valley, Massachusetts: p. 48–60.

Stern, T. W., and Rose, H. J., Jr., 1961, New results from lead-alpha age measurements: Am. Mineralogist, v. 46, nos. 5–6, p. 606–612.

Stewart, D. P., 1961, The glacial geology of Vermont: Vermont Geol. Survey Bull. 19, 124 p.

Stille, Hans, 1940a, Zur Frage der Herkunft der Magmen: Preuss. Akad. Wiss. Abh. 1939, no. 19, 31 p.

——1940b, Einführung in den Bau Amerikas: Berlin, Gebrüder Borntraeger, 717 p.

——1950, Der "subsequente" Magmatismus: Deutsche Akad. Wiss. Berlin, Kl. Math. u. Naturw. 1950, Abh. Geotektonik 3, 25 p.

Stockwell, C. H., 1961, Structural provinces, orogenies, and time classification of rocks of the Canadian Precambrian Shield, *in* Reports, Part 2 *of* Age determinations by the Geological Survey of Canada: Canada Geol. Survey Paper 61–17, p. 108–118.

——1962, A tectonic map of the Canadian Shield, *in* The tectonics of the Canadian Shield: Royal Soc. Canada Spec. Pub. 4, p. 6–15.

——1963a, Second report on structural provinces, orogenies, and time-classification of rocks of the Canadian Precambrian Shield, *in* Geological studies, Part 2 *of* Age determinations and geological studies: Canada Geol. Survey Paper 62–17, p. 123–133.

——1963b, Third report on structural provinces, orogenies, and time-classification of rocks of the Canadian Precambrian Shield, *in* Geological studies, Part 2 *of* Age determinations and geological studies: Canada Geol. Survey Paper 63–17, p. 125–131.

——1964, Fourth report on structural provinces, orogenies, and time-classification of rocks of the Canadian Precambrian Shield, *in* Geological studies, Part 2 *of* Age determinations and geological studies: Canada Geol. Survey Paper 64–17, p. 1–21.

——1965a, Structural trends in Canadian Shield: Am. Assoc. Petroleum Geologists Bull., v. 49, no. 7, p. 887–893.

——1965b, Tectonic map of the Canadian Shield: Canada Geol. Survey Map 4–1965.

——1968, Geochronology of stratified rocks of the Canadian Shield: Canadian Jour. Earth Sci., v. 5, no. 3, p. 693–698.

Stone, S. W., and Dennis, J. G., 1964, The geology of the Milton quadrangle, Vermont: Vermont Geol. Survey Bull. 26, 79 p.

Swift, C. M., Jr., 1966, Geology of the southeast portion of the Averill quadrangle, New Hampshire: New Hampshire Dept. Resources and Econ. Devel. Quad. Rept.

Taylor, E. F., 1964, Vermont has oil, gas possibilities: World Oil, v. 158, no. 7, p. 142, 144, 146.

Thayer, T. P., 1966, Serpentinization considered as a constant-volume metasomatic process: Am. Mineralogist, v. 51, nos. 5–6, p. 685–710.

Theokritoff, George, 1964, Taconic stratigraphy in northern Washington County, New York: Geol. Soc. America Bull., v. 75, no. 3, p. 171–190.

——1968, Cambrian biogeography and biostratigraphy in New England, p. 9–22, *in* Zen, E-an, White, W. S., Hadley, J. J., and Thompson, J. B., Jr., *Editors*, Studies of Appalachian Geology: Northern and Maritime: New York, Interscience Publishers, 475 p.

Thomas, J. J., 1968, The detection of low-angle thrust surfaces by mesostructural analysis, as applied in western Vermont: Kansas Univ., Lawrence, Ph.D. dissert., 91 p.

Thompson, G. A., and Talwani, Manik, 1964, Crustal structure from Pacific Basin to central Nevada: Jour. Geophys. Research, v. 69, no. 22, p. 4813–4837.

Thompson, J. B., Jr., 1952, Southern Vermont, *in* Geol. Soc. America Guidebook for field trips in New England, Field Trip 1, Geology of the Appalachian Highlands of east-central New York, southern Vermont, and southern New Hampshire: p. 14–23, 38–41.

——1954, Structural geology of the Skitchewaug Mountain area, Claremont quadrangle, Vermont-New Hampshire, *in* New England Intercollegiate Geol. Conf. Guidebook 46th Ann. Mtg., Hanover, N. H., Oct. 1954: p. 37–41.

——1955, The thermodynamic basis for the mineral facies concept: Am. Jour. Sci., v. 253, no. 2, p. 65–103.

——1957, The graphical analysis of mineral assemblages in pelitic schists: Am. Mineralogist, v. 42, nos. 11–12, p. 842–858.

——1959, Stratigraphy and structure in the Vermont Valley and the eastern Taconics between Clarendon and Dorset (Vermont), *in* New England Intercollegiate Geol. Conf. Guidebook 51st Ann. Mtg., Rutland, Vt., Oct. 1959, Stratigraphy and structure of west-central Vermont and adjacent New York: p. 71–87.

Thompson, J. B., Jr., and Norton, S. A., 1968, Paleozoic regional metamorphism in New England and adjacent areas, p. 319–327, *in* Zen, E-an, White, W. S., Hadley, J. J., and Thompson, J. B., Jr., *Editors*, Studies of Appalachian Geology: Northern and maritime: New York, Interscience Publishers, 475 p.

Thompson, J. B., Jr., Robinson, Peter, Clifford, T. N., and Trask, N. J., Jr., 1968, Nappes and gneiss domes in west-central New England, p. 203–218, *in* Zen, E-an, White, W. S., Hadley, J. J., and Thompson, J. B., Jr., *Editors*, Studies of Appalachian Geology: Northern and Maritime: New York, Intersicience Publishers, 475 p.

Thompson, J. B., Jr., and Rosenfeld, J. L., 1951, Tectonics of a mantled gneiss dome in southeastern Vermont (Abs.): Geol. Soc. America Bull., v. 62, no. 12, Part 2, p. 1484–1485.

Tilton, G. R., Davis, G. L., Wetherill, G. W., and Aldrich, L. T., 1957, Isotopic ages of zircon from granites and pegmatites: Am. Geophys. Union Trans., v. 38, no. 3, p. 360–371.

Tilton, G. R., Wetherill, G. W., Davis, G. L., and Bass, M. N., 1960, 1000-million-year-old minerals from the eastern United States and Canada: Jour. Geophys. Research, v. 65, no. 12, p. 4173–4179.

Trask, N. J., Jr., and Thompson, J. B., Jr., 1967, Stratigraphy and structure of the Skitchewaug nappe in the Bernardston area, Massachusetts and adjacent New Hampshire and Vermont, *in* New England Intercollegiate Geol. Conf. Guidebook 59th Ann. Mtg., Amherst, Mass., Oct. 1967, Field trips in the Connecticut Valley, Massachusetts: p. 129–142.

Traverse, Alfred, 1955, Pollen analysis of the Brandon lignite of Vermont: U.S. Bur. Mines Rept. Inv. 5151, 107 p.

Traverse, Alfred, and Barghoorn, E. S., 1953, Micropaleontology of the Brandon lignite, an early Tertiary coal in central Vermont: Jour. Paleontology, v. 27, no. 2, p. 289–293.

Turner, F. J., 1948, Mineralogical and structural evolution of the metamorphic rocks: Geol. Soc. America Mem. 30, 342 p.

Uchupi, Elazar, 1966, Structural framework of the Gulf of Maine: Jour. Geophys. Research, v. 71, no. 12, p. 3013–3028.

Ulrich, E. O., and Schuchert, Charles, 1902, Paleozoic seas and barriers in eastern North America: New York State Mus. Bull. 52, p. 633–663.

van Bemmelen, R. W., 1963, Geotektonische Stockwerke, eine relativistische Hypothese der Geotektonik: Geol. Gesell. Wien. Mitt. 1962, v. 55, p. 209–232.

——1965, Der gegenwärtige Stand der Undationstheorie: Geol. Gesell. Wien. Mitt., v. 57, no. 2, p. 379–399.

Van Hise, C. R., 1896, Principles of North American pre-Cambrian geology: U.S. Geol. Survey Ann. Rept. 16, Part 1, p. 571–843.

Voight, Barry, 1965, Structural relationships of the Sudbury nappe to the subjacent Middlebury synclinorium and superjacent Taconic allochthon in west-central Vermont (Abs.): Geol. Soc. America Spec. Paper 82, p. 214–215.

Walton, M. S., and de Waard, D., 1963, Orogenic evolution of the Precambrian in the Adirondack highlands, a new synthesis: Koninkl. Nederlandse Akad. Wetensch. Proc., ser. B, v. 66, no. 3, p. 98–106.

Wanless, R. K., Stevens, R. D., Lachance, G. R., and Edmunds, C. M., 1968, K-Ar isotopic ages, Rept. 8, *in* Age determinations and geological studies: Canada Geol. Survey Paper 67–2, Part A, 141 p.

Wanless, R. K., Stevens, R. D., Lachance, G. R., and Rimsaite, J. Y. H., 1966, K-Ar isotopic ages, Rept. 6, *in* Age determinations and geological studies: Canada Geol. Survey Paper 65–17, 101 p.

Watson, Janet, 1964, Conditions in the metamorphic Caledonides during the period of late-orogenic cooling: Geol. Mag., v. 101, no. 5, p. 457–465.

Weiss, L. E., 1958, Structural analysis of the basement system at Turoka, Kenya: Great Britain, Overseas Geology and Mineral Resources, v. 7, nos. 1 and 2, p. 3–35, 123–153.

Welby, C. W., 1961, Bedrock geology of the Central Champlain Valley of Vermont: Vermont Geol. Survey Bull. 14, 296 p.

White, W. S., 1946, Rock-bursts in the granite quarries at Barre, Vermont: U.S. Geol. Survey Circ. 13, 23 p.

——1949, Cleavage in east-central Vermont: Am. Geophys. Union Trans., v. 30, no. 4, p. 587–594.

——1959, A revision of the lower Paleozoic stratigraphy in eastern Vermont; a discussion: Jour. Geology, v. 67, no. 5, p. 577–581.

White, W. S., and Billings, M. P., 1951, Geology of the Woodsville quadrangle, Vermont-New Hampshire: Geol. Soc. America Bull., v. 62, no. 6, p. 647–696.

White, W. S., and Jahns, R. H., 1950, Structure of central and east-central Vermont: Jour. Geology, v. 58, no. 3, p. 179–220.

Wiesnet, D. R., and Clark, T. H., 1966, The bedrock structure of Covey Hill and vicinity, northern New York and southern Quebec, *in* Geological Survey research 1966: U.S. Geol. Survey Prof. Paper 550-D, p. D35-D38.

Williams, C. R., and Billings, M. P., 1938, Petrology and structure of the Franconia quadrangle, New Hampshire: Geol Soc. America Bull., v. 49, no. 7, p. 1011–1043.

Willis, Bailey, and Willis, Robin, 1934, Geologic structures, 3d ed., revised: New York, McGraw-Hill Book Co., Inc., 544 p.

Wilson, J. T., 1966, Did the Atlantic close and then re-open?: Nature, v. 211, no. 5050, p. 676–681.

Winkler, H. G. F., 1967, Petrogenesis of metamorphic rocks (revised 2d ed.): New York, Springer-Verlag, 237 p.

Woodland, B. G., 1963, A petrographic study of thermally metamorphosed pelitic rocks in the Burke area, northeastern Vermont: Am. Jour. Sci., v. 261, no. 5, p. 354–375.

——1965, The geology of the Burke quadrangle, Vermont: Vermont Geol. Survey Bull. 28, 151 p.

Woodward, H. P., 1957, Structural elements of northeastern Appalachians: Am. Assoc. Petroleum Geologists Bull., v. 41, no. 7, p. 1429–1440.

Wyllie, P. J., 1963, The nature of the Mohorovicic discontinuity, a compromise: Jour. Geophys. Research, v. 68, no. 15, p. 4611–4619.

——1965, A modification of the geosyncline and tectogene hypothesis: Geol. Mag., v. 102, no. 3, p. 231–245.

Wynne-Edwards, H. R., 1964, The Grenville Province and its tectonic significance: Geol. Assoc. Canada Proc., v. 15, Part 2, p. 53–67.

Young, G. M., 1968, Miogeoclines (miogeosynclines) in space and time: A discussion: Jour. Geology, v. 76, no. 1, p. 116–119.

Zartman, R. E., Brock, M. R., Heyl, A. V., and Thomas, H. H., 1967, K-Ar and Rb-Sr ages of some alkalic intrusive rocks from central and eastern United States: Am. Jour. Sci., v. 265, no. 10, p. 848–870.

Zartman, R. E., Hurley, P. M., Krueger, H. W., and Giletti, B. J., 1970, A Permian-disturbance of radiometric ages in New England—its occurrence and causes: Geol. Soc. America Bull. (in press).

Zartman, R. E., Snyder, G. L., Stern, T. W., Marvin, R. F., and Bucknam, R. C., 1965, Implications of new radiometric ages in eastern Connecticut and Massachusetts, in Geological Survey research 1965: U.S. Geol. Survey Prof. Paper 525-D, p. D1-D10.

Zen, E-an, 1960, Metamorphism of lower Paleozoic rocks in the vicinity of the Taconic Range in west-central Vermont: Am. Jour. Sci., v. 45, nos. 1–2, p. 129–175.

——1961, Stratigraphy and structure at the north end of the Taconic Range in west-central Vermont: Geol. Soc. America Bull., v. 72, no. 2, p. 293–338.

——1963a, Age and classification of some Taconic stratigraphic units on the Centennial geologic map of Vermont—A discussion: Am. Jour. Sci., v. 261, no. 1, p. 92–94.

——1963b, Structural relations in the southern Taconic region; an interpretation, *in* Geol. Soc. America Guidebook Field Trip 3, Stratigraphy, structure, sedimentation, and paleontology of the southern Taconic region, eastern New York: p. 1–4.

——1963c, Components, phases, and criteria of chemical equilibrium in rocks: Am. Jour. Sci., v. 261, no. 10, p. 929–942.

——1964a, Taconic stratigraphic names—definitions and synonymies: U.S. Geol. Survey Bull. 1174, 95 p.

——1964b, Stratigraphy and structure of a portion of the Castleton quadrangle, Vermont: Vermont Geol. Survey Bull. 25, 70 p.

——1967, Time and space relationships of the Taconic allochthon and autochthon: Geol. Soc. America Spec. Paper 97, 107 p.

——1968, Nature of the Ordovician orogeny in the Taconic area, p. 129–139, *in* Zen, E-an, White, W. S., Hadley, J. J., and Thompson, J. B., Jr., *Editors*, Studies of Appalachian Geology: Northern and Maritime: New York, Interscience Publishers, 475 p.

Zen, E-an, and Hartshorn, J. H., 1966, Geologic map of the Bashbish quadrangle, Massachusetts, Connecticut, and New York: U.S. Geol. Survey Geol. Quad. Map GQ-507.

Zwart, H. J., 1967, The duality of orogenic belts: Geologie en Mijnbouw, Jaarg. 44, p. 283–309.

PUBLICATION AUTHORIZED BY THE DIRECTOR, U.S. GEOLOGICAL SURVEY

Appendix A: Inner Zone of the Orthogeosyncline in Southeastern New England

The inner (axial) zone of the orthogeosyncline, apparently in the belt of the Acadian Merrimack synclinorium in south-central New Hampshire, east-central Massachusetts, and eastern Connecticut, is probably the southeastern source of sediments and also the root of early recumbent folds, both of which were transported northwestward into the region of the present study. The nature of the inner zone is not clear, other than that it is probably all eugeo-synclinal, inasmuch as modern fieldwork which is being done with an awareness developed during the study of less obscure outer belts of the orthogeosyncline has not been completed. Nevertheless, interregional studies indicate that the terrane of the Merrimack synclinorium in south-central Massachusetts and Connecticut covers, or perhaps actually includes, the transition from the "Pacific" faunal realm to the "Atlantic" or "Acado-Baltic" realm, of the lower Paleozoic. (*See* Shaw, 1961; Theokritoff, 1968, p. 19–20.) Tectonic interpretation of this transition will perhaps be far reaching and may eventually help explain the regional sedimentational and tectonic relations and also their accompanying magmatic and metamorphic features (Poole, 1967, p. 45–46; Wilson, 1966).

Pelites and semipelites that overlap geanticlines in the northwestern part of the orthogeosyncline and that are separated from the craton by miogeosynclinal limestones and dolomites require a southeastern source of sediments. These pelites and semipelites include rocks in the Merrimack synclinorium in New Hampshire that have been assigned to the Devonian and Silurian(?) Littleton Formation, but that (at least in their western exposures) include the Ordovician Partridge Formation (Thompson, 1954, p. 36, 38,; Thompson and others, 1968, p. 206, Pl. 15–1a, 15–1b). The Partridge Formation and underlying rocks are traced southward into lower Paleozoic units that fill most of the southern extension of the Merrimack synclinorium in Massachusetts and Connecticut (Dixon and Lundgren, 1968; Dixon and Shaw, 1965; Eaton and Rosenfeld, 1960, Table 1; Lundgren, 1962; Figs. 1, 2; Robinson, 1967b, p. 19; Thompson and others, 1968, Pl. 15–1a, 15–1b; Zartman and others, 1965, Fig. 1). Therefore, much of the rock in the Merrimack synclinorium, hitherto considered middle Paleozoic, may very likely be lower Paleozoic and a possible source of lower and middle Paleozoic sediments. Such a source would suggest a northeast-trending belt of early to middle Paleozoic intrageosynclinal uplift (Pl. 3), although no stratigraphic evidence for such an uplift (stratigraphic convergence and unconformity) has been reported.

The early recumbent folds, mostly northwest-facing in northwestern New England and adjacent Quebec, disregard the geanticlines of this region much as did the northwestward transport of sediments just discussed, and therefore also suggest a belt to the southeast of early to middle Paleozoic uplift, possibly including lateral compression. A southeastward-facing early recumbent fold reported in southeastern New England (Dixon and others, 1963; Dixon and Lundgren, 1968, p. 221–225) indicates southeastward movement apparently from a line or belt of parting with the northwest-facing folds (Robinson, 1967a, Fig. 4).

A ground-surface slope away from the vicinity of the root zones of the recumbent folds (as well as from the source of sediments) seems unavoidable because the rocks did not have the strength to form the folds (some more than 15 miles from synclinal root to anticlinal hinge; *see* Thompson and others, 1968, Pl. 15–1a, 15–1b) by other than gravity flow (Hubbert, 1937, p. 1498–1499). The ground-surface slopes may have been produced simply by rise of buoyant rocks beneath to form an intrageosynclinal undation (van Bemmelen, 1965, p. 395–396), in which case the root zone was initially about parallel to the slopes, or they may have been produced by the squeezing and up-welling of rocks as a result of bench-vise shortening of the cross section of the geosynclinal trough under lateral compression (Thompson and others, 1968, p. 216). According to the latter more traditional interpretation the root zones were initially subvertical and at large angles to the ground-surface slopes analogous to the angle between the bottom and top surfaces of some glaciers in the zone of upflow of bottom ice within the lip of a cirque. Late (climactic Acadian) folding has changed the initial attitudes of roots of the folds making it difficult to determine by which process they were formed.

The belt of early and middle Paleozoic intrageosynclinal uplift postulated in the terrane of the Merrimack synclinorium coincides fairly closely with the staurolite and sillimanite zones of regional metamorphism. Isograds are closely spaced adjoining the terrane, which spacing indicates a steep thermal gradient and rapid heat flow; their surfaces also dip steeply and are inverted in the recumbent folds (Thompson and Norton, 1968, p. 325; Thompson and others, 1968, p. 215–216). These relationships were perhaps brought about by geosynclinal subsidence and geothermal heating of rocks that were then buoyed or squeezed up, flowed laterally, and were quenched before heat flow was equalized, then afterward steeply tilted in the late folds of the synclinorium. (*See* Chinner, 1966, p. 178.) The concordant calc-alkalic plutons (Mount Clough and Cardigan plutons and perhaps Fitchburg pluton—Billings, 1955) that are exposed at the borders of the synclinorium may have formed at the time of heating and metamorphism (Thompson and Norton, 1968, p. 325; Thompson and others, 1968, p. 216) and could have contributed to the buoyancy. The closeness of at least one of the plutons (Mount Clough) to the metamorphic granitic rocks in the domal anticlines ("Oliverian domes"), and its compositional similarity to the metamorphic granitic rocks, (*see* Billings and Keevil, 1946, Fig. 3, p. 806, 817) suggests that the magma was produced by melting of some of the metamorphic granites of depth. (*See* Rutland, 1966, p. 186.)

Appendix B: Chemical Affinities of the Initial Magmatic Features

Predominantly tholeiitic affinities of the initial magmatic features are shown by the results of 28 chemical analyses (only six of which have been published to date (Billings and Wilson, 1964, Table 4, p. 34–35; Fairbairn, 1933, Table 1, p. 556) of mostly extrusive greenstones and amphibolites (Table 4). Rocks of 19 analyses fall in the tholeiitic field of MacDonald and Katsura's (1964, p. 87) alkali-silica (Na_2O+K_2O/SiO_2) diagram (Fig. 5); rocks of 9 analyses fall in the alkalic field. Thirteen of those that plot in the tholeiitic field also fall in the tholeiitic field of Kuno's (1960, p. 137, Fig. 10) Al_2O_3/Na_2O+K_2O diagram; three are near the mutual boundaries of the tholeiitic-high-alumina-, and alkali-basalt fields; two are near the boundary between the tholeiitic and high-alumina basalt fields, and one is near the boundary between the high-alumina- and alkali-basalt fields. Of the nine that plot in MacDonald and Katsura's alkalic field, two fall in Kuno's alkalic field, three are near the mutual boundaries between the tholeiitic-, high alumina-, and alkali-basalt fields, two are near the boundaries between the high-alumina- and alkali-basalt fields, and two contain too little silica to be represented in Kuno's diagrams. The most alkalic of these rocks are to the northwest in the eugeosynclinal zone, nearest the craton, and the most aluminous are to the southeast, where felsic volcanics are most abundant and where concordant calc-alkalic granitic plutons are also common.

Recalculated to sums of 100 percent, with H_2O and CO_2 discarded, the rocks of only three of the 28 analyses (numbers 1, 2, and 24) fall in the alkalic field of MacDonald and Katsura, and one (number 1) is within the alkalic field of Kuno.

Oceanic theoleiitic affinities are indicated by an SiO_2 content that varies from 48 to 50.5 percent in 11 of the 19 samples that are tholeiitic in MacDonald's and Katsura's breakdown, and by a K_2O content of commonly less than 0.5 percent in 16 of these samples, following the criterion of Engel and others (1965, p. 728, Fig. 3; p. 730, Fig. 5). Oceanic tholeiitic affinities, especially affinity with the tholeiites of mid-ocean ridges (Engel and others, 1965, p. 723, 725) is indicated by a TiO_2 content of 1.5 percent or less in 25 out of a total of the 28 samples of initial magmatic rocks analyzed. This suggests that among the 25, those showing apparent alkalic affinities have been post-magmatically altered, perhaps spilitized or metamorphically differentiated, thereby raising the ratio of $Na_2O + K_2O$ to SiO_2.

TABLE 4. MAJOR CHEMICAL CONSTITUENTS AND CLASSIFICATIONS OF MAFIC VOLCANIC AND HYPABYSSAL ROCKS IN NORTHWESTERN NEW ENGLAND AND ADJACENT QUEBEC

Constituents	1	2	3	4	5	6	7	8	9	10
SiO_2	47.2	42.6	45.7	45.0	49.0	48.93	48.4	48.9	51.0	41.7
Al_2O_3	15.0	15.4	14.0	13.5	14.7	14.88	15.2	14.3	14.2	18.9
Fe_2O_3	3.0	4.11	4.4	4.9	3.9	2.1	3.7	2.9	2.9	5.5
FeO	10.0	10.94	8.1	8.4	6.4	8.55	6.6	8.1	6.9	7.0
MgO	7.5	9.3	10.4	10.9	8.2	7.99	7.0	8.9	8.4	9.8
CaO	6.5	10.57	11.9	11.3	11.5	11.51	13.2	10.0	10.7	7.3
Na_2O	3.1	2.11	1.7	1.9	2.6	2.02	2.3	2.9	2.9	1.2
K_2O	.80	.6	.51	.25	.25	.15	.07	.08	.09	.03
H_2O^-	.12	.04			.09	.03				.12
H_2O^+	4.2	2.76	1.7	2.2	2.4	2.48	1.8	2.7	2.2	5.6
TiO_2	1.9	1.15	.76	1.0	.82	.99	1.0	1.0	.89	.94
P_2O_5	.31	.01	.11	.16	.10	.08	.15	.13	.12	.10
MnO	.24	.28	.20	.24	.16	.21	.20	.29	.20	.14
CO_2	.05	.01	<.05	<.05	.05	.01	<.05	<.05	<.05	.43
Total	99.92	99.88	99.53	99.80	100.17	99.93	99.67	100.25	100.55	98.76
Classifications: MacDonald and Katsura (1964)	Alkalic	Alkalic	Tholeiitic	Tholeiitic	Tholeiitic	Tholeiitic	Tholeiitic	Tholeiitic	Tholeiitic	Alkalic
Kuno (1960)	Alkalic		Tholeiitic	Tholeiitic	Tholeiitic	Tholeiitic	{Tholeiitic High-alumina Alkalic}	Tholeiitic	Tholeiitic	
Engel, Engel and Havens (1965) SiO_2 = 48–50.5%					Oceanic	Oceanic	Oceanic	Oceanic		
K_2O < 0.5%				Oceanic	Oceanic	Oceanic	Oceanic	Oceanic	Oceanic	

Rapid rock analyses 1, 5, 12, 17, 19, 20, 22, 23 by P. L. D. Elmore, S. D. Botts, I. H. Barlow, and G. W. Chloe; 3, 4, 7, 8, 9 by P. L. D. Elmore, K. E. White, S. D. Botts; 10, 11, 13, 14, 15, 16, 18 by P. L. D. Elmore, S. D. Botts, and G. W. Chloe. Standard rock analyses 2, 6 by E. J. Tomasi and F. H. Neuerburg; 21, 25, 27, 28 by F. A. Gonyer; 24 by Pease; and 26 by C. Kahn.

Constituents	11	12	13	14	15	16	17	18	19	20
SiO_2	48.2	49.2	49.4	46.7	45.4	47.8	49.8	48.7	49.9	50.9
Al_2O_3	14.1	15.9	15.4	14.6	15.1	15.6	16.0	13.4	14.5	16.5
Fe_2O_3	4.7	2.4	8.1	4.4	5.5	6.6	2.9	5.2	5.8	3.6
FeO	6.9	6.3	6.6	8.0	4.8	6.0	6.4	3.2	5.2	7.1
MgO	5.9	7.9	6.1	7.9	6.0	5.7	8.1	6.8	6.9	6.9
CaO	11.9	12.6	6.0	10.3	13.1	10.9	6.6	14.7	11.7	7.1
Na_2O	2.4	2.1	2.1	2.0	2.4	3.0	4.3	2.1	1.7	3.1
K_2O	.45	.14	.08	.32	.20	.51	.10	.17	.12	.03
H_2O^-	.06	.12	.05	.06	.04	.04	.16	.04	.08	.10
H_2O^+	2.3	2.5	4.2	2.8	2.8	1.4	3.6	1.6	2.9	3.9
TiO_2	1.5	.85	1.4	1.1	.90	1.0	.98	.92	1.0	1.1
P_2O_5	.18	.10	.17	.14	.11	.12	.11	.14	.13	.11
MnO	.19	.18	.18	.19	.19	.18	.24	.12	.18	.20
CO_2	.47	.05	.15	.71	3.2	.08	.81	1.8	.05	.05
Total	99.25	100.34	99.93	99.22	99.74	98.93	100.10	98.89	100.16	100.69
Classifications:										
MacDonald and Katsura:	Tholeiitic	Tholeiitic	Tholeiitic	Tholeiitic	Alkalic	Alkalic	Alkalic	Tholeiitic	Tholeiitic	Tholeiitic
Kuno:	Tholeiitic	Tholeiitic	Tholeiitic	Tholeiitic	Tholeiitic / High-alumina / Alkalic	Tholeiitic / High-alumina / Alkalic	Alkalic	Tholeiitic	Tholeiitic	Tholeiitic / High-alumina
Engel, Engel and Havens:										
SiO_2 = 48–50.5%	Oceanic	Oceanic	Oceanic					Oceanic	Oceanic	
K_2O < 0.5%	Oceanic	Oceanic	Oceanic	Oceanic				Oceanic	Oceanic	Oceanic

TABLE 4 (continued)

Constituents	21	22	23	24	25	26	27	28
SiO_2	49.44	50.4	44.9	46.55	53.44	50.91	52.40	51.64
Al_2O_3	15.92	15.9	15.8	19.26	17.8	16.0	18.18	17.62
Fe_2O_3	2.33	1.7	3.0	2.58	3.11	1.17		1.14
FeO	8.11	7.1	7.7	9.73	6.18	8.81	5.59	7.8
MgO	6.10	7.6	6.9	6.67	6.24	6.85	6.26	7.74
CaO	8.20	8.5	9.4	9.07	5.4	9.99	4.64	6.44
Na_2O	3.06	4.1	2.2	3.31	3.1	2.27	5.04	4.52
K_2O	.50	.04	.34	.09	.26	.69	.58	.25
H_2O^-	3.40	.07	.10	2.39		.05		
H_2O^+		2.9	4.6		4.0	.97	3.84	1.29
TiO_2	1.43	.88	2.0	.52	.51	1.68	.90	1.33
P_2O_5		.10	.26		.11	.20	.19	.09
MnO	.10	.19	.20	.25	.12	.21	.10	.12
CO_2	1.31	.54	3.1			.07	2.13	
Total	99.90	100.02	100.50	100.42	100.27	99.87	99.85	99.98
Classifications:								
MacDonald and Katsura:	Tholeiitic	Tholeiitic	Alkalic	Alkalic	Tholeiitic	Tholeiitic	Alkalic	Tholeiitic
Kuno:	Tholeiitic High-alumina Alkalic	Tholeiitic High-alumina Alkalic	Tholeiitic High-alumina Alkalic	High-alumina Alkalic	Tholeiitic High-alumina	Tholeiitic	High-alumina Alkalic	High-alumina Alkalic
Engel, Engel and Havens:								
$SiO_2 = 48$–50.5%	Oceanic	Oceanic						
$K_2O < 0.5\%$		Oceanic			Oceanic			Oceanic

Stratigraphic units and localities from which samples were collected

Cambrian(?) Pinnacle Formation:

1. Tibbit Hill Volcanic Member (greenstone), 1.0 mile N 5° E of Fletcher, Vermont.

Cambrian(?) and Lower Cambrian Hazens Notch Formation, Belvidere Mountain Amphibolite Member, east slope of Belvidere Mountain, Eden and Lowell, Vermont:

2. Coarse amphibolite.
3. Do.
4. Do.
5. Do.
6. Fine amphibolite.
7. Do.
8. Do.
9. Do.

Lower Ordovician Stowe Formation, mafic volcanic and hypabyssal rocks:

10. Volcanic greenstone, 0.25 mi. N 65° W of Big Falls, Troy, Vermont.
11. Volcanic greenstone, 0.1 mi. S 55° E of Stowe, Vermont.
12. Dike metagabbro, 1.35 mi. S 35° W of Lowell, Vermont.
13. Volcanic greenstone, Stowe Pinnacle, 3 mi. S 40° E of Stowe, Vermont.
14. Volcanic amphibolite, 0.8 mi. S 23° W of Scrag Mountain, Waitsfield, Vermont.
15. Volcanic greenstone, 0.8 mi. N 50° E of Burnt Mountain, Northfield, Vermont.
16. Volcanic amphibolite, 2.0 mi. N 80° E of Morrisville, Vermont.
17. Volcanic greenstone, 2.0 mi. S 82° E of Bean Mountain, Eden, Vermont.
18. Volcanic greenstone, 0.3 mi. west of East Granville, Vermont.
19. Dike greenstone, 2.1 mi. N 59° E of Rice Mountain, Roxbury, Vermont.

Middle (and Lower?) Ordovician Missisquoi Formation of Doll and others (1961) and correlatives, mafic volcanic and hypabyssal rocks:

20. Dike greenstone, 3.5 mi. S 87° E of Eden Mills, Vermont.
21. Volcanic greenstone, Potten, Quebec (Bolton Group, uralite—Cooke, 1950, p. 79–81; Fairbairn, 1933, p. 556).
22. Volcanic greenstone, Coburn Hill Volcanic Member of the Missisquoi Formation, 0.7 mi. N 20° E of Summit Siding, Newport, Vermont.
23. Dike greenstone, 1.9 mi. N 17° E of Montpelier Junction, Vermont.

Middle Ordovician Ammonoosuc Volcanics and correlatives (Billings and Wilson, 1964, p. 34–35):

24. Chlorite schist (Orfordville Formation), Hanover, New Hampshire.
25. Chlorite-epidote schists, 1.71 mi. N 70° W of Walker Mountain, Littleton, New Hampshire.
26. Amphibolite, 0.2 mi. S 65° E of Bowman, Randolph, New Hampshire.

Upper Silurian(?) and Lower Devonian Littleton Formation (Billings, 1937, p. 556):

27. Volcanic greenstone, 0.8 mi. north of Walker Mountain, Littleton, New Hampshire.
28. Volcanic amphibolite, 0.65 mi S 45° W of Streeter Pond, Lisbon, New Hampshire.

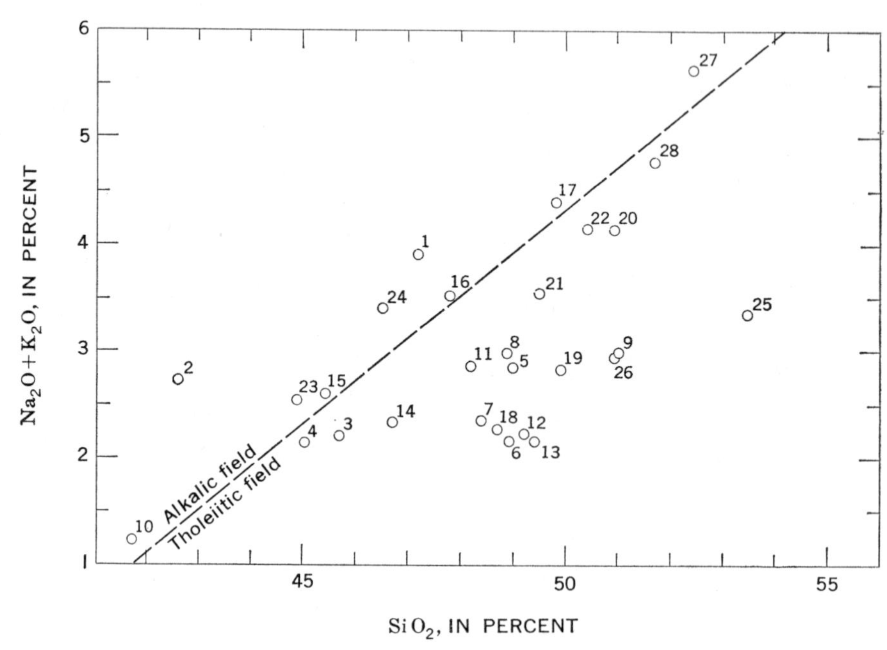

Figure 5. Alkali: silica diagram of mafic volcanic and hypabyssal rocks in northwestern New England and adjacent Quebec (modified *from* MacDonald and Katsura, 1964, p. 87, Fig. 1).

Appendix C: General Interpretation of the Metamorphic Processes

The regional metamorphic rocks discussed in this paper belong mostly to the kyanite-sillimanite type facies series of Miyashiro (1961, p. 278–279), which is synonymous with Barrow's metamorphic zonation in the Grampian Highlands of Scotland as discussed by Harker (1939, p. 185–187). The low-pressure intermediate group of Miyashiro (1961, p. 282–283) is also represented, but is found chiefly bordering discordant (post-kinematic) granitic plutons, and is thus less of a regional feature and postdates the tectonic features of the orthogeosyncline. (*See also* Albee, 1968, p. 329.) Various recent discussions and illustrations of regional metamorphism pertinent to the author's studies are provided by Bailey and others (1964, Fig. 21, p. 110), Bateman and Eaton, 1967, p. 1415–1416, Billings (1956, p. 136–140), Guidotti (1965, p. 106–110), Hamilton (1963, p. 88–89), Hamilton and Myers, 1967, p. C13-C15, C23-C24, Hollingworth (1962), Miyashiro (1961, p. 284–287; 1967), Shimazu (1961), Thompson (1955, p. 98–99; 1957, p. 842–845), Winkler (1967, p. 84–130, 173–191), and Zwart (1967).

The author's interpretation of the metamorphic processes, without specific reference to the regional tectonic relations, is as follows: The vertical component of the overall pressure gradient caused by the load of overlying rocks increased with depth. Where the rocks had enough strength, however, the overall gradient was probably somewhat lessened locally by components of tectonic overpressure, such as those that cause folding, transmitted subhorizontally in the more shallow rocks. The temperatures also increased with depth, and were apparently at first maintained near "normal" thermal gradient despite geosynclinal subsidence and deposition, which was slow enough to allow for heat flow from subjacent and adjacent rocks into the cool new bedded rocks. The hydrous aluminum silicates—chlorite, muscovite, and biotite, and the foliation for which they are responsible—are probably initial products of this "normal" thermal gradient working hand in hand with the pressure gradient produced by the load of overlying rocks in the geosyncline. This is suggested by the distribution of chlorite, muscovite, and biotite, which shows little correlation with the distribution of igneous rocks whose magmatic heat would have steepened the thermal gradient, or with that of folds that might imply tectonic overpressure. The nonhydrous aluminum silicates, garnet and kyanite, could partly be products of somewhat less than "normal" thermal gradients and relatively higher load and fluid pressure gradients, produced where subsidence and concomitant deposition of the geosynclinal rocks was too rapid for the flow of heat into them to keep pace, and the outward flow of water was encouraged by the desiccating effect of immediately subjacent dry basement rocks. Garnet and kyanite are possibly in part, however, products of postgeosynclinal rise of geoisotherms in anticlinal tracts. Kyanite veins found locally, in otherwise lower grade rock, probably reflect high water pressure in combination with relatively low temperature and lithostatic pressure within veins from which water did not readily escape. (*See also* Newton, 1966, p. 1224–1225.)

A steeper than "normal" thermal gradient appears to explain the nonhydrous aluminum silicates, staurolite and sillimanite, as in the Buchan terrane of Scotland or low-pressure intermediate group of Miyashiro (1961, p. 283), previously discussed, principally because the distribution of these minerals corresponds rudely with that of intrusive igneous rocks (Billings, 1956, p. 140–141). There are, however, extensive areas where one or the other of these minerals is found, in which the igneous rocks fail to crop out. Possibly the necessary heating is to be ascribed to the magma of concordant granitic plutons that dip beneath or have been eroded from above the metamorphic rocks, especially those bearing sillimanite; although at least the latter seems unlikely inasmuch as the metamorphic rocks dropped down from above, during post-metamorphic cauldron subsidence connected with emplacement of discordant alkalic plutons, are of chlorite rather than sillimanite grade (*see* Billings, 1956, p. 29). If the plutons were never extensive in the vicinity of the sillimanite-bearing rocks, a steeper than "normal" thermal gradient may be attributed to relatively rapid upward flow of heat from deeper geosynclinal rocks (*see* Bateman and Eaton, 1967, p. 1416) or from the Earth's mantle (*see* Thompson and Talwani, 1964, p. 4834; Wyllie, 1963, p. 4616–4617) in that part of the region characterized especially by sillimanite. Perhaps such heat was primarily responsible not only for regional metamorphism but, with ultrametamorphism, for generation of the magma of the concordant plutons (*see* Thompson and Norton, 1968, p. 325; Thompson and others, 1968, p. 208), and through these for major uplift within the eugeosynclinal zone (*see* van Bemmelen, 1963, p. 225; Wyllie, 1965). Thickness and deformation of the bedded rocks containing staurolite and sillimanite may be no more than the thickness and deformation of the bedded rocks containing garnet and kyanite, hence pressures and "normal" geothermal heating responsible for the staurolite and sillimanite are probably no greater (*see also* Albee, 1965, p. 297–298; Clark and others, 1957, p. 638), and seemingly not as significant as abnormally rapid upward heat flow.

Appendix D: Metasomatic Processes Connected With the Ultramafic Rocks

The metasomatic adjustments between the ultramafic rocks and the bedded wallrocks in which they are emplaced involve exchange of material across their mutual contacts.

Rodingite and serpentine-chlorite rock (including free carbon) were produced in the wallrocks by simple reconstitution and by exchange of material with the ultramafites. These processes were brought about perhaps partly by residual magmatic heat in the dunite and peridotite, but chiefly during incipient metamorphism by lowering of the partial pressures of water and carbon dioxide in the wallrock near the contacts (inward transfer of water and oxygen and expulsion of hydrogen) as desiccating and reducing effects of dunite and peridotite, bringing about replacement of the latter by serpentinite (Cady and others, 1963, p. 46; Chidester, 1968, p. 352–353; *see also* Aniruddha, 1968; Olsen, 1961, p. 345; Thayer, 1966, p. 704; Thompson, 1955, p. 98–99). Steatite, blackwall chlorite and biotite rock, and albite porphyroblast rock likewise replaced the bedded wallrock as a part of an exchange across the contacts. The other part of this exchange within the ultramafic rock was replacement of serpentinite by steatite, talc-carbonate rock, and magnesite-quartz rock. This exchange is considered an accompaniment of continued regional metamorphism of the bedded rocks, during which carbon dioxide activity increased in the bedded rocks causing carbonatization of serpentinite within the ultramafic bodies. This carbonatization, in turn, among other effects triggered exchange of water and magnesium in the serpentinite for silicon in the wallrock (Chidester, 1962, p. 91–124; 1968, p. 351–352; Jahns, 1967, p. 157–160).

The increased carbon dioxide activity inferred in the bedded rocks could be attributed to release of oxygen from iron oxides in sediments during metamorphism followed by its recombination with carbon of nearby carbonaceous phyllites to form carbon dioxide. This latter process seems fairly clear inasmuch as "cleaned up" phyllites are common; in these phyllites iron oxide is found only in the form of magnetite and the only carbon occurs as relics in albite porphyroblasts (Cady and others, 1962; *see also* Zen, 1963c, p. 934–936).

Appendix E: Age of the Rocks of the Grenville Province

Lower Precambrian eugeosynclinal rocks of the Superior province of the Canadian Shield (Keewatin Series) apparently extend southeastward into the Grenville province (Osborne and Morin, 1962, Table 1, p. 133). Also, Rb-Sr whole-rock isochrons of granitic rocks emplaced in bedded rocks on both sides of the transition (Grenville Front) between the rocks of the Grenville and Superior provinces, show early Precambrian (Rb-Sr isochron) ages of approximately 2.4 b.y. (Grant, 1964; Grant and others, 1965) and another such whole-rock isochron of granitic rocks 25 miles southeast of the transition shows a similar age (Davis and others, 1967, p. 383).

The type Grenville Series has been tentatively assigned to the middle Precambrian (Stockwell, 1964, p. 18–21; 1965b) because miogeosynclinal rocks have been traced from known middle Precambrian (Huronian and Kaniapiskau), in the Southern and Churchill provinces of the Canadian Shield, into the Grenville province. Alternatively, a late Precambrian age for the type Grenville Series has been suggested on the circumstantial evidence that rocks like those of the type Grenville and like some of its more assured stratigraphic correlatives seem widely absent from the Grenville province, perhaps because after deposition on the structural and metamorphic terrane of the Hudsonian orogeny at the end of the middle Precambrian the Grenville rocks were largely eroded before the billion-year cooling event (Wynne-Edwards, 1964, p. 54–68; *see also* Appleyard, 1965, p. 51, 55). Possibly, middle or upper Precambrian rocks, or both, were deposited only in southeastern areas never having lapped northwestward on early to middle Precambrian terrane now exposed in the type Grenville area and in the northwestern part of the Grenville province. Such an interpretation is consistent with the southeastward offlap of Paleozoic geosynclinal deposits in the Appalachian belt.

Upper Precambrian bedded rocks have been recognized in at least one area, which borders the Gulf of St. Lawrence near the southeast edge of the Grenville province (Grenier, 1957, p. 62–65; Lowdon, 1961, p. 76; Stockwell, 1965b). These rocks are likely correlatives of those in the Precambrian basement in the Adirondack Mountains and beneath the orthogeosynclinal Paleozoic rocks of

New England and southeastern Quebec, because of both their southeastern geographic position and their late Precambrian age.

Pegmatites and granites a little more than a billion years old mark the culmination of regional metamorphism and intrusive igneous activity ending in the billion-year cooling event (Engel and Engel, 1953, p. 1044; Hills and Gast, 1964; Silver, 1964; *see also* Faul and others, 1963, p. 3, and Tilton and others, 1960). The dates and origins of plutons of anorthosite and related rocks in the Adirondacks and the Grenville province are more controversial. The latter rocks are variously interpreted as magmatic intrusives that have been emplaced in and are therefore younger than the Grenville Series (Buddington, 1939, p. 201–221; 1960; Osborne and Morin, 1962, p. 124–126; Stockwell, 1964, p. 20; 1965b), as palingenetic derivatives of the Grenville (Kranck, 1961, p. 314–317), or as an early basement upon which the Grenville sediments were deposited and which was reactivated when the Grenville was deformed (de Waard and Walton, 1967; Walton and de Waard, 1963; Wynne-Edwards, 1964, p. 62–65). Their age, depending upon one's interpretation of their origin, has been made more or less explicitly both early and late Precambrian (Osborne and Morin, 1962, p. 125, 131, 135; Stockwell, 1964, p. 20).

Appendix F: Previously Unpublished Radiometric Age Determinations of the U.S. Geological Survey

Lab No.	Field No.	Sample Locality	Rock and Unit	Age[1] K-Ar	Age[1] Rb-Sr
173	VT-16	44°06′37″ N. 73°00′30″ W.	Metagraywacke, Cambrian(?) Pinnacle Fm.	390(B)	350 ± 20(B)
174	VT-20	44°21′58″ N. 72°40′44″ W.	Garnet-kyanite schist, Lower Ordovician Stowe Fm.	430(M)	
285	VT-14	43°58′01″ N. 73°01′48″ W.	Pegmatite, Precambrian Mount Holly Complex	770(M)	
286	VT-15	44°05′43″ N. 72°59′00″ W.	Biotite schist, Precambrian Mount Holly Complex	410(B)	
287	VT-17	44°07′50″ N. 73°01′00″ W.	Phyllite, Cambrian(?) Fairfield Pond Mbr.,[2] Underhill Fm.	360(R)	
288	VT-18	44°21′28″ N. 73°05′15″ W.	Phyllite, Cambrian(?) Fairfield Pond Mbr.,[2] Underhill Fm.	385(R)	
514B	GI-62–7	44°38′43″ N. 73°20′55″ W.	Lamprophyre, Mesozoic dike	136(B)[3]	120 ± 80(B)

[1] Letter symbols: B = biotite, M = muscovite, R = whole rock
[2] Doll and others (1961)
[3] Zartman and others (1967, p. 862)

All but sample 514B were collected jointly by W. M. Cady, A. H. Chidester, and Henry Faul; sample 514B was collected by A. V. Heyl.

Analytical data for K-Ar determinations:

LAB. No.	PERCENT K_2O	$^*Ar^{40}$(PPM)	$^*Ar^{40}/K^{40}$	PERCENT $^*Ar^{40}$	CALCULATED AGE (M.Y.)
173	7.67	0.196	0.0255	96	390(B)
174	8.24	.236	.0286	98	430(M)
285	10.48	.592	.0557	99	770(M)
286	9.59	.259	.0267	98	410(B)
287	5.67	.134	.0233	98	360(R)
288	6.76	.172	.0251	98	385(R)
514B	7.94	.0662	.00822	92	136(B)

Ages are calculated by means of the following equations and constants:

$$T = \frac{1}{\lambda e + \lambda \beta} \ln \left[\frac{^*Ar^{40}}{K^{40}} \left(\frac{\lambda e + \lambda B}{\lambda e} \right) + 1 \right]$$

Decay constants $K^{40}\lambda\beta = 4.72 \times 10^{-10}$/yr, $\lambda e = 0.585 \times 10^{-10}$/yr

Abundance $K^{40} = 1.22 \times 10^{-4}$gm/gm K

* Radiogenic isotope

Potassium determinations were made with a Perkin-Elmer flame photometer with lithium as internal standard. Analysts were H. H. Thomas, R. F. Marvin, P. L. D. Elmore, and H. Smith.

Argon determinations were made by isotope dilution methods. Analysts were H. H. Thomas and R. F. Marvin.

The overall analytical error is approximately ±5 percent of the listed age value. Most ages have been rounded to the nearest 5 m.y.

Analytical data for Rb-Sr age determinations:

LAB No.	Rb^{87} (PPM)	NORMAL SR (PPM)	$^*Sr^{87}$ (PPM)	$^*Sr^{87}/Rb^{87}$	$^*Sr^{87}/Sr^{87}$ TOTAL	CALCULATED AGE (M.Y.)
173	105	13.5	0.541	0.00514	0.364	350 ± 20(B)
514B	92.5	369.0	.166	.00179	.007	120 ± 80(B)

Ages are calculated by means of the following equations and constants:

$$T = \frac{1}{\lambda\beta} \ln \left[\frac{^*Sr^{87}}{Rb^{87}} + 1 \right]$$

Decay constant $Rb^{87}\lambda\beta = 1.47 \times 10^{-11}/yr$

Abundance $Rb^{87} = 0.283$ *gm/gm Rb*

*Radiogenic isotope

Initial Sr^{87}/Sr^{86} assumed to be 0.703 in sample 173, determined from coexisting apatite to be 0.7040 in sample 514B.

Rubidium and strontium determinations were made by mass spectrometric isotope dilution methods. Analysts were F. Walthall, W. D. Long, and R. E. Zartman. Possible analytical error is shown with the age.

Appendix G: Hybrid Late Ordovician and Silurian K-Ar Ages

Apparent Ordovician and Silurian K-Ar dates may reasonably be interpreted as hybrids of Cambrian and Devonian regional metamorphic events, despite objections to the contrary.

The principal objections may first be stated, as follows:
(1) The metamorphosed bedded rocks that provided these K-Ar values are well enough reconstituted to have been completely degassed in a final episode of folding at the time of the Taconic disturbance; micas were used in the measurements and very likely would not have produced anomalously old K-Ar dates. (2) There is no evidence for a stratigraphic break, hence little probability of a very early episode of erosional unloading and resetting of K-Ar systems within the Cambrian and Ordovician in areas that have provided the values interpreted as hybrid. (3) The K-Ar values correspond with the Taconic stratigraphic break. (4) Values corresponding to the time span, 414 m.y. to 460 m.y., are widely distributed areally and include in addition some Rb-Sr whole-rock ages. (5) The micas dated include some that form polygonal arches (that is, they are not bent) in minor folds and that less commonly are parallel to the axial-plane cleavage of the folds. Hence they are new micas, formed at the close of or after folding, whose K-Ar ages are interpreted as true Taconic ages not possibly produced by hybridization of mica formed before folding.

These objections are answerable as follows: (1) The stratigraphic and structural relations of the Green Mountain-Sutton Mountain anticlinorium indicate that it was folded in the Acadian orogeny—the subsidiary folds on the southeast limb of the anticlinorium, at and northeast of Lake Memphremagog, Quebec, involve paleontologically well established Silurian and Devonian strata. Moreover, Silurian and Lower Devonian strata are included in the east limb of the axial anticline of the anticlinorium in southern Vermont. Hence it appears that the final and also dominant episode of folding was sometime after the Lower Devonian strata were deposited, and that degassing and full resetting of the K-Ar systems was probably not finished until during or after the Acadian orogeny. One of the samples of mica measured (VT-20, Appendix F) was, at the time of the Taconic disturbance, overlain by 25,000 feet of Ordovician rock that remained undissected by erosion (Cady, 1956, map and sections) and possibly blocked critical cooling and resetting of the K-Ar systems of the sample both during and after the Taconic disturbance. (2) Stratigraphic breaks that mark episodes of erosion and probably concomitant critical cooling within the Cambrian and Ordovician may not be everywhere apparent, especially in geosynclinal belts. Where folding took place in the geosynclinal rocks, chiefly recumbent isoclinal folds were formed; their limbs and axial surfaces were parallel to the sides and bottoms of the geosynclines and thus were parallel to comparable features in other folds, or simply to undeformed bedding, although separated from them by stratigraphic breaks. (3) The time spread of the K-Ar dates, 460 m.y. to 414 m.y. ago, is mainly Middle and Late Ordovician through at least Middle and possibly Late Silurian; therefore, there is little coincidence with the Taconic disturbance. (4) The areally widespread values corresponding to this age span quite possibly include some dates that are real and reflect degassing events actually connected with the Taconic disturbance—where they are beyond the range of regional metamorphism connected with the Acadian orogeny. The Rb-Sr whole-rock values in this age span may also include some values that reflect the Taconic disturbance. (5) The critical new micas form polygonal arches in, and are parallel to, axial-plane cleavage of minor folds that are genetic parts of the major late folds. The latter have flexed Silurian and Lower Devonian as well as Cambrian and Ordovician strata. Hence the new micas are not possibly Taconic, but appeared near the close of or after the Early Devonian, probably during the Acadian orogeny. The hybrid K-Ar ages obtained are possibly explained by mixture of the new micas with old micas that had been formed and degassed and their K-Ar systems reset in pre-Taconic episodes and survived degassing during or soon after the Acadian, and by survival of some of the material of the old micas at the nuclei of the new. Or perhaps the argon of the new micas may be anomalously high because some was absorbed from the old micas during Acadian deformation under high gas pressures that, before erosion after the Acadian orogeny, prevailed at depth in the geosynclinal pile. Such a phenomenon has been observed in minerals that originated in the Earth's mantle (Lovering and Richards, 1964; Zartman and others, 1967, p. 862–865), but it has never been adequately documented in minerals from crustal rocks. (*See* Moorbath, 1967, p. 119–120.)

Appendix H: Rb-Sr Age Determinations of the Tibbit Hill Volcanic Member of the Cambrian (?) Pinnacle Formation

M. A. Lanphere, who performed Rb-Sr measurements on sericite phyllite from the Cambrian(?) Tibbit Hill Volcanic Member (as used by Doll and others, 1961) of the Pinnacle Formation, stated that the date of 494 ± 20 m.y. ago should be considered a minimum value and that its failure to agree with the field relationships may be attributed to several possible factors. First, in the absence of a suite of samples from which to construct an Rb-Sr isochron it was necessary to assume an initial isotopic composition of the strontium, which he said, if wrong by 0.5 percent, could introduce an error of 30 m.y. in the calculated age. Or, the measured age may be low due to metamorphic effects. He also stated that it is very important to have a representative fragment of the total rock sample, which was not attempted.

The author collected three samples totaling 270 pounds that were intended originally for determination of Pb/alpha age of zircon; this determination proved to be impossible because the zircon is only in traces and enough for radiometric analysis could not be separated. A piece from one of the samples that contained a relatively small amount of metamorphically segregated rock material was then selected for the Rb-Sr determination. Information concerning this piece is as follows:

Field No.: CA-179-C. Collected and examined petrographically by W. M. Cady.

Locality: 1.3 mi. WNW of Rochelle, Quebec, Orford Sheet (west) of National Topographic Series; roadcut south of highway from Rochelle to Waterloo; lat 45° 24′ 40″N., long. 72° 25′ 30″ W.

Local relationships: Specimen from the base of a fresh roadcut taken 6 feet below glaciated surface of bedrock and about 350 feet WNW of easternmost bedrock exposure in the cut.

Geology: Gray phyllitic rock mottled with a few white spots and transected by axial-plane cleavage that strikes N.28°E. and dips 87°W. Microscopically the rock is a granular mosaic, chiefly of grains of orthoclase and quartz, replacing and interstitial to which is sericite (muscovite?) oriented parallel to the cleavage. The white spots are thin laminar quartz segregations, also parallel to the axial plane cleavage. Rounded, presumably corroded, microscopic phenocrysts of potash feldspar are set in the groundmass. Sericite and calcite replace both phenocrysts and groundmass. The grains of the groundmass range in size from 0.005 to 0.025 mm; the phenocrysts are 0.5 to 2.0 mm in diameter. The estimated mode is as follows:

	Groundmass	Phenocrysts	Composite
Orthoclase	48	1	49
Microcline	..	1	1
Microcline-perthite	..	1	1
Quartz	29	..	29
Sericite	10	tr	10
Magnetite	5	tr	5
Calcite	3	..	3
Sphene-leucoxene	2	..	2
Apatite	tr	..	tr
	97	3	100

Preparation and radiometric analysis: Done by M. A. Lanphere, using facilities of the Division of Geological Sciences, California Institute of Technology. A piece weighing about 400 gm was crushed to −80 mesh and split to 1.34 gm for analysis. The strontium composition was measured directly as a portion of this split, and concentrations of strontium and rubidium were measured on other portions by means of isotope dilution techniques.
Analytical data:

	10^{-6}gm/gm	10^{-6}moles/gm
Sr^{87*}	0.1597	0.001837
Normal Sr	18.16	.2072
Rb	81.86	.9567

(Sr^{87}/Sr^{86}) measured = 0.9472
(Sr^{87}/Sr^{86}) common = 0.708
Rb^{87} = 1.39 — 10^{-11} yr^{-1}

Appendix I: Intrusive and Stratigraphic Relations of Radiometrically Dated Granitic Dike Rocks and of Ultramafic Rocks in Which They Are Emplaced, in the Thetford Area, Quebec

The ultramafic and mafic rocks have been considered to postdate the Beauceville Formation as well as the underlying Caldwell Formation, for the reason that they are "intruded along a zone which follows the contact of the Caldwell and Beauceville rocks" (Riordon, 1954, p. 3; 1957, p. 391; 1962, p. 311; *see also* Wanless and others, 1966, p. 79). The granitic rocks are emplaced only in the ultramafic and mafic rocks and have been considered to be of about the same age (Riordon, 1962, p. 311).

The ± 480 m.y. K-Ar dates of the granitic rocks have thus been interpreted as the minimum radiometric age of the Beauceville, and of upper Middle Ordovician rocks worldwide, which is 35 m.y. greater than the estimate of 445 m.y. indicated in Kulp's (1961) time scale or possibly 20 m.y. greater than the 460 m.y. indicated in Holmes (1959) time scale (Poole and others, 1963, p. 1063–1064). It has been further argued that the 480 ± m.y. dates reflect recrystallization that accompanied deformation of the Beauceville, inasmuch as they are from muscovites in cataclastic and recrystallized granitic rock (Poole and others, 1964; Wanless and others, 1966, p. 79). It is acknowledged in an alternative explanation (Poole and others, 1963, p. 1064; *see also* Wanless and others, 1966, p. 79–80) that "if the ultramafic and contained muscovitic granitic rocks actually became coherent bodies well below the present surface much earlier than the Middle Ordovician and then were emplaced together as solids after the Middle Ordovician, the K-Ar dates would indicate an undetermined age of formation at depth modified perhaps by incomplete degassing of muscovite."

Commenting on these alternative interpretations, A. G. Smith (*in* Harland and others, 1964, p. 367) remarked, "The restriction of the stocks and dykes to the ultramafic rocks, their cataclastic texture, and poor agreement with the Middle Ordovician dates . . . suggest that the second interpretation is correct"—

specifically, "that the ages are those of micas formed much earlier below their present structural level, incompletely degassed during later intrusion as solid masses." Rickard (1964b, p. 913) suggested that the Beauceville lies unconformably on gabbro and the ultramafic rocks, implying that these igneous rocks intrude only the St. Daniel and Caldwell Formations, beneath the unconformity. He further suggested, "perhaps the date of 477–481 m.y. is the date of pre-Taconic orogeny."

Author Index

Subject Index